BRITISH MAPS OF COLONIAL AMERICA

PUBLISHED FOR THE HERMON DUNLAP SMITH CENTER
FOR THE HISTORY OF CARTOGRAPHY
THE NEWBERRY LIBRARY

DAVID WOODWARD, SERIES EDITOR

PREVIOUSLY PUBLISHED

MAPS: *A Historical Survey of Their Study and Collecting*
by R. A. SKELTON (1972)

William P. Cumming

BRITISH MAPS OF COLONIAL AMERICA

A PLAN of the
TOWN and HARBOUR of
BOSTON.
and the Country adjacent with the
Road from Boston to Concord.
Shewing the Place of the late Engage-
ment between the Kings Troops & the
Provincials, together with the several
Encampments of both Armies in and
about Boston. 19th April 1775.

THE KENNETH NEBENZAHL, JR., LECTURES IN THE HISTORY OF CARTOGRAPHY AT THE NEWBERRY LIBRARY

THE UNIVERSITY OF CHICAGO PRESS | CHICAGO & LONDON

THE UNIVERSITY OF CHICAGO PRESS, CHICAGO 60637
THE UNIVERSITY OF CHICAGO PRESS, LTD., LONDON
© 1974 by The University of Chicago
All rights reserved. Published 1974
Printed in the United States of America
International Standard Book Number: 0–226–12362–6
Library of Congress Catalog Card Number: 73–84190

Title page: A Plan of the Town and Harbour of Boston . . . (1775)
[manuscript].
Courtesy of the Duke of Northumberland, Alnwick.

WILLIAM P. CUMMING is Irvin Professor Emeritus, Davidson College, Davidson, North Carolina. His many publications include *The Southeast in Early Maps* and *The Discovery of North America* (with R. A. Skelton and D. B. Quinn).
[1974]

526.80973
C971

To Louis C. Karpinski
Lawrence C. Wroth
R. A. Skelton

Valued mentors
and
well-remembered friends

Contents

Figures

Preface

The second series of the Kenneth Nebenzahl, Jr., Lectures in the History of Cartography was given in the Fellows' Lounge of the Newberry Library in April 1970. The lectures, entitled "The British Cartography of Eighteenth-Century North America," originally illustrated by two hundred slides and accompanied by an exhibition of the maps of the period in the Newberry Library, have been modified for publication, with the addition of two appendixes, notes, and a bibliographical essay.

In this study of British colonial cartography it is necessary to explain its geographical and chronological limits and to define the term British. The survey concerns itself with a great period of English mapmaking in North America and its emphasis is primarily on the eighteenth century. It is helpful to look at earlier maps, however, to understand what was achieved in the eighteenth century. The essays discuss mapmakers, their methods, and the historical background giving rise to their production as well as the maps themselves. Since British cartography, brilliantly active during the Revolutionary War years, slowed almost to a stop (at least south of the Canadian provinces) after the treaty of peace of 1783, this study ends there. British cartography excludes the product of continental mapmakers; surveys of foreigners living in the British colonial dominions or employed in British government service, such as the Swiss Captain John Collet, the Dutch Bernard Romans, and the German Gerard De Brahm, are not excluded. Neither are the maps of the colonial settlers who were British subjects and so regarded themselves.

This survey could not have been made without the previous investigations of many scholars whose work I gratefully acknowledge. As anyone who wants to make use of early maps soon finds, however, the gaps in knowledge are serious and the desiderata many; in every direction the limits of available information are reached with depressing suddenness. Basic cartobibliographical lists are lacking for maps of regions like New England; or for general and local maps of a particular province like New York or Pennsylvania. Such lists are diminished in value and effectiveness if they do not include manuscript as well as printed maps. The individual contributions of important mapmakers like Des Barres and of half a dozen others discussed in these pages need thorough examination that they have not received. A few major maps deserve intensive study of their sources and relationship to other maps, of their contributors, and of the historical, geographical, and ethnic information that they contain. Fortunately, this has been done for two or three Pennsylvania maps; the students of several disciplines would profit from such a study of Douglass's *New England,* Mitchell's *British and French Dominions,* or Stuart-Purcell's *Southeast.* Yet insufficiently known are the methods and instruments used by the surveyors and mapmakers of colonial America and the training in the craft given to officers who made the maps, often beautifully designed and executed, in the wars. Journals, newspapers and magazines, and military and naval audit files have been insufficiently tapped for such information.

I wish to thank Mr. and Mrs. Kenneth Nebenzahl for making these lectures possible and for their hospitality to Mrs. Cumming and me during our visits to Chicago. I thank Dr. Lawrence W. Towner, Director of the Newberry Library, for his professional cooperation and both Dr. and Mrs. Towner for their many personal kindnesses. Dr. David Woodward, Curator of Maps of the

Newberry Library, has been unfailing in his help during the preparation of these lectures, in organizing the fine map exhibit that accompanied them, and especially in his editorial care in preparing the manuscript for publication.

I am especially indebted to Miss Jeannette Black, whose wealth of knowledge I have repeatedly called on with profit and whose critical judgment has enabled me to avoid pitfalls; and to Dr. Helen Wallis, who shared with me exciting discoveries in the examination of the Bernard Collection and of the Royal United Services Institution Collection. I thank the Duke of Northumberland for his hospitality at Alnwick Castle and his gracious permission for *American Heritage* to photograph the Percy maps. I also thank Dr. and Mrs. J. G. C. Spencer Bernard for their kindness in receiving us at Nether Winchendon House and allowing us to study the Bernard Collection. Professor T. R. Smith of the University of Kansas, Professor Coolie Verner of the University of British Columbia, Professor Louis De Vorsey of the University of Georgia, Dr. R. A. Skelton, the late Superintendent of the Map Room, the British Museum, and Mrs. Newman Hall of Washington, D. C., have with exceptional generosity sent me unpublished manuscripts that I have used in preparing these lectures.

Mr. P. A. Penfold of the Public Record Office, Mr. T. A. Corfe of the British Museum Map Room, Mr. R. W. Stephenson of the Library of Congress Geography and Map Division, and Mr. J. Aubrey of the Department of Special Collections, the Newberry Library, have given me time and aid whenever I have called upon them.

Finally, my wife, Mrs. Elizabeth C. Cumming, has been my indefatigable research assistant and helpful critic during the preparation of these lectures.

WILLIAM P. CUMMING

Mapping the Southern British Colonies

Expanding Knowledge of the Interior and the Settlement of the South

The subject of this second series of the Kenneth Nebenzahl, Jr., Lectures in the History of Cartography is the achievement of the British in the mapping of their American colonies in the eighteenth century. Much of the field is unploughed territory, and a mere short-title list of the pertinent British maps produced during the period would run to hundreds of pages. Obviously, my treatment must be highly selective. If, therefore, a map of this period is passed over or receives summary treatment, this may be the result of my desire to give consideration to newly discovered maps or to treat a topic so far unconsidered. What I do hope to convey is an idea of the copiousness and increasing excellence of the British cartography of their American colonies and to illustrate some of the reasons why this mapping was so important to them. The need to record new knowledge of the fast-moving frontier and the expanding settlements, to assert the British position in boundary disputes, to facilitate trade, to chart the coasts and the routes of campaigns: these were among the most powerful motives. Needing a division for four lectures, I have chosen somewhat arbitrarily to treat, in turn, the cartography of the south, of the north, of the coast, and of the wars. It is obvious, however, that there will be

a certain amount of overlapping: coasts extend both north and south and so do conflicts; some important maps cover the whole area; and specialized kinds of cartography are better treated topically rather than regionally.

It is appropriate to begin this examination with the southern colonies, since the first known map of any part of North America by an Englishman based on firsthand information is a sketch map of the North Carolina Outer Banks area, sent back to England in 1585 from Raleigh's Roanoke colony.[1] It was probably made by the painter and later governor of the colony, John White, or by the brilliant young mathematician, Thomas Harriot, whose more careful surveys of the region with White are found in White's manuscript drawings in the British Museum and the engraved map, published by Theodore de Bry in 1590, that accompanied Harriot's *A brief and true report of the new founde land of Virginia*, sometimes called the beginning of American literature.[2]

Sixty years later, John Farrer, long an official of the Virginia Company, designed a fascinating map that was engraved and published by John Stephenson in 1651. Farrer's map, which appeared in several works and in a number of editions, shows an unusual combination of knowledge concerning the coast and gross misconceptions of the interior.[3] The Pacific Ocean, Farrer states, is only ten days' march west of the head of the James River beyond the Blue Ridge. Throughout most of the seventeenth century, English merchants and Virginians still hoped for a short overland route, making the Virginia colony middleman for commerce between the Orient and Europe. The extremely rare first state of the plate of Farrer's map has a strait in the top right corner leading to the Pacific from the Hudson River. In later states it is blocked by an isthmus (see fig. 1). The imaginary geography so current in seventeenth-century maps gradually gave way to expanding knowledge in the eighteenth century. On the Farrer map, north is at the right; it was not until the eighteenth century that orientation with north at the top was established. The legend in the upper part of the map reads:

1

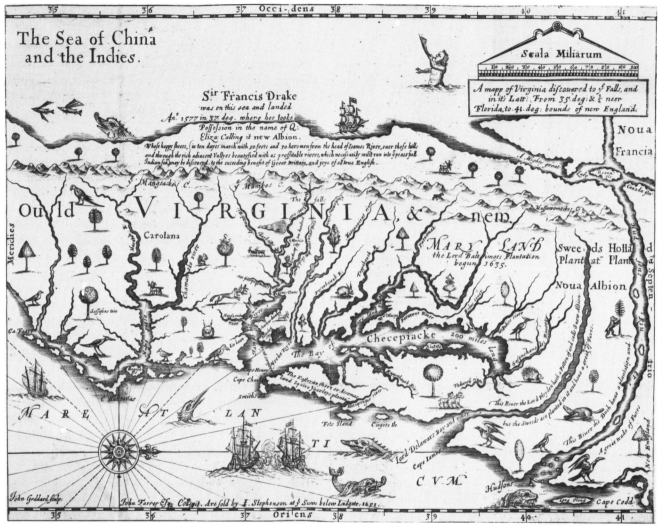

Figure 1 John Farrer, *A mapp of Virginia discouered to ye Falls* (1651), in: Edward Williams, *Virgo Triumphans* (London, 1650).

Courtesy of the Ayer Collection, the Newberry Library.

2

Sir Francis Drake was on this sea and landed Ano. 1577 in 37. deg. where hee tooke Possession in the name of Q: Eliza: Calling it *new Albion*. Whose happy shoers, (in ten dayes march with 50. foote and 30. hors·men from the head of Ieames River, ouer those hills and through the rich adiacent Vallyes beautyfied with as proffitable rivers, which necessarily must run into yt peacefull Indian sea,) may be discovrd to the exceeding benefit of Great Brittain, and joye of al true English.

Drake actually made that landing on the coast of California, probably slightly north of San Francisco.

THE LORDS PROPRIETORS' MAPS OF CAROLINA

In 1663 Charles II rewarded eight courtiers who had supported his return to the throne by giving them, with great generosity in lands that he did not own in the first place, all the region between Virginia and Florida and westward from the Atlantic to the Pacific. John Ogilby made a map of Carolina, usually called the First Lords Proprietors' map, based largely on reports and a map given him by John Locke, philosopher, author of the Fundamental Constitutions of Carolina, and secretary to Lord Ashley, one of the Lords Proprietors (fig. 2). On this map, now in the Public Record Office, Locke designated landmarks by names of the Lords Proprietors, such as Albemarle Sound, Clarendon County, Craven County, the Ashley and Cooper rivers. To fill the interior, Ogilby used a map made by a young German, John Lederer, who had been sent on an exploratory expedition by Governor Sir William Berkeley of Virginia to look beyond the Blue Ridge with hope of seeing the Pacific on the other side.[4] Lederer said that Piedmont North Carolina was a savanna under water several months of the year, that he saw a great lake (nonexistent), and that he returned to Virginia by the pine barrens of North Carolina, which he called the Arenosa Desert. Forty years ago, walking down Charing Cross Road, I saw a copy of Ogilby's map in an old print shop window. I had never thought of my home in the Piedmont as being under water half the year, with a nearby Arenosa Desert one hundred and

twenty-five miles long. Many of Lederer's misconceptions continued to appear on maps as late as the middle of the eighteenth century. It was my attempt to find out the reasons for these geographical vagaries, suggested to me by Professor L. C. Karpinski, that began my interest in the history of cartography.

Two other maps of the seventeenth century illustrate the beginning of a trend that becomes important in the eighteenth-century British cartography of the South: the recording of names and locations of individual plantation owners, especially of the new landed aristocracy of Virginia and Charles Town.

In the ten years since the preparation of Ogilby's map, interior exploration and coastal navigation had contributed new and much more accurate delineation as shown in 1682 on the Second Lords Proprietors' map, by Joel Gascoyne (fig. 3).[5] All the Lederer detail is omitted. The Gascoyne map in the British Museum is an earlier state than any in this country; those in American libraries have added trees, animals, and geographical features to the interior, and have changed the location of the Westo Indian settlement and the size and location of a lake in "A Pleasant Valley" situated on a branch of the Combahee River. Since no lake ever existed in that region, the change on the plate is a mystery. Traces of revision are visible on the later state of the plate here reproduced; there is no change in the cartouche, however, and the crudely erased lines in the lower half of the imprint are already deleted in the earlier British Museum copy. So we shall never know what was engraved in this space unless we find an earlier state of the map.

Gascoyne's map is based upon the work of the Surveyor General of Carolina, Maurice Mathews. A few years later (about two years before his death in 1687), Mathews made a great manuscript chart of Carolina, now in the British Museum, which gave the location of some 250 landowners around Charles Town. In 1695 John Thornton and Robert Morden used Mathews's great chart as the basis for the most detailed printed map of any large area in the Southeast since White's map one hundred years before.[6]

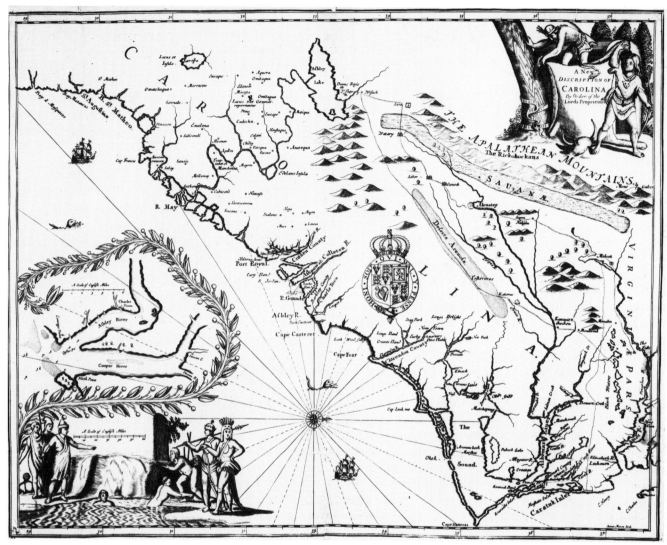

Figure 2 John Ogilby, *A New Discription of Carolina,* in: John
Ogilby, *America* (London, ca. 1672). Courtesy of the Newberry
Library.

4

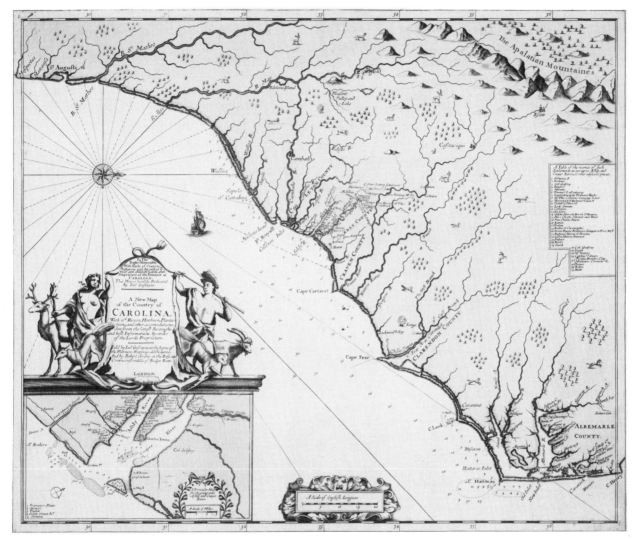

Figure 3 Joel Gascoyne, *A New Map of the Country of Carolina* ([1682]), in: Joel Gascoyne, *A True Description of Carolina* (London, [1682]). Courtesy of the Yale University Library Map Collection.

5

EARLY EIGHTEENTH-CENTURY CARTOGRAPHY

Lawrence C. Wroth has called the Edward Crisp map of 1711 "one of the glories of Anglo-American cartography."[7] The large central map, surrounded by insets, is still based on Maurice Mathews's great chart but is enlarged in area, with new plantation owners added by the surveyor John Love (fig. 4). To digress here for a moment, a significant and revealing picture of contemporary surveying methods is given by John Love in his *Geodaesia* (1688), a popular work on surveying that went through twelve editions.[8] It should be emphasized that surveying methods in the colonies were crude: triangulation surveys, already being made on a large scale in Europe by the Cassinis, Jean Picard, and others in the seventeenth century, were probably not used in the southern colonies before the middle of the eighteenth century. In an appendix to the second edition (1715), John Love gave the following advice "as . . . how to lay out new lands in America." "Shewing farther," he writes,

How to survey by the Chain only . . . What has been said concerning Measuring a Field or taking an Angle by the Chain only, . . . may as well be applied to a Pole or Rod cut out of the Hedge and divided into 100 equal Parts; and indeed you may altogether as well, and much quicker, do it with a Rod than a Chain . . . But then you must take care that your Rod be straight . . . You may have, I suppose, in *Crooked-Lane,* a Rod made to shoot, one Part into another like a Fishing-rod, to be used as a Cane, in the Heal whereof may be a small Compass. Which alone is instrument enough to survey any Piece of This Earth, be it a Mannor or larger; And if so, what need is there of a Horseload of Brass Circles, and Semicircles, heavy Ballsockets, Wooden Tables and Frames, and 3-legged Staffs, *Cum multis aliis,* unless to amuse the ignorant Countryman, to make him more freely pay the Surveyor.

With Love's *Geodaesia* as one of the more popular eighteenth-century surveying manuals in America, it is not surprising that boundary disputes and quitrent problems became frequent. In the second half of the century such methods gave way to truly expert surveying and draftsmanship.

The main section of the Crisp map shows the settlements around Charles Town with the names of the owners of the plantations situated on the banks of the tidal rivers. The bottom left, extending southward to the Savannah River past Port Royal and Hilton Head, shows Edisto Island, almost uninhabited in Mathews's time, now well settled along the waterfronts. Another inset is the Spanish town of Saint Augustine, included without justification by Charles II within the boundaries of Carolina. Still another is the much-copied plan of Charles Town; below it is a poor chart of the entire South Atlantic coast. The southern part of Florida is shown as an archipelago, a conception of Florida found on some maps even after the middle of the century. The eighteenth century was also not without its geographical misconceptions.

The most important inset is one of the Southeast made by Captain Thomas Nairne (fig. 5). By about 1710 the Charles Town empire builders, Nairne and Colonel "Tuscarora Jack" Barnwell, with other expansionists, had pushed south past the Spanish to the Everglades in Florida and west to the Mississippi River in bitter but successful competition with the French for the Indian trade. Captain Nairne's inset is the first map of the Southeast based upon British colonial exploration as far west as the Mississippi River.

In fact, larger and more serious matters than local land surveying were beginning to trouble the colonies, as the Nairne inset shows; international boundary disputes were appearing on maps. France as well as England had its empire builders, and rival claims initiated a cartographic war of considerable interest. Guillaume Delisle made a series of maps showing French claims that agitated persons high in office in England. His 1703 map of North America shows French expansion in the Mississippi Valley and delimits British territories to the east of the Appalachian mountains. A branch of the Mississippi on this map, which almost reaches "Checagou," is one of the early representations of the Chicago

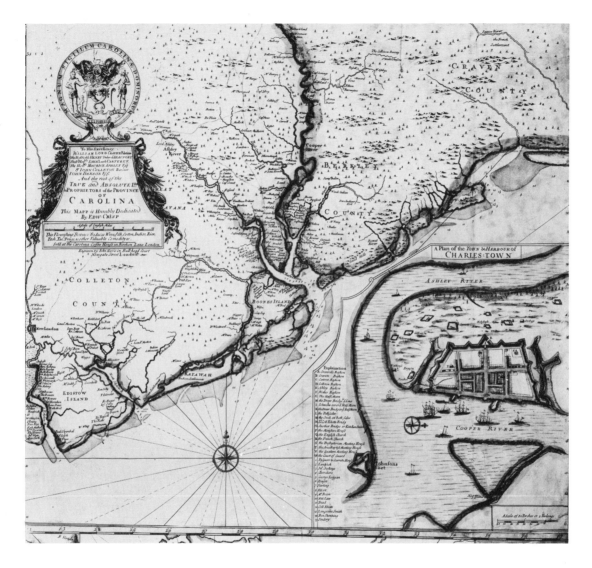

Figure 4 Edward Crisp [Charleston, South Carolina, and vicinity, detail] from *A Compleat Description of the Province of Carolina* (London, [1711]) [separately published]. Courtesy of the Library of Congress.

7

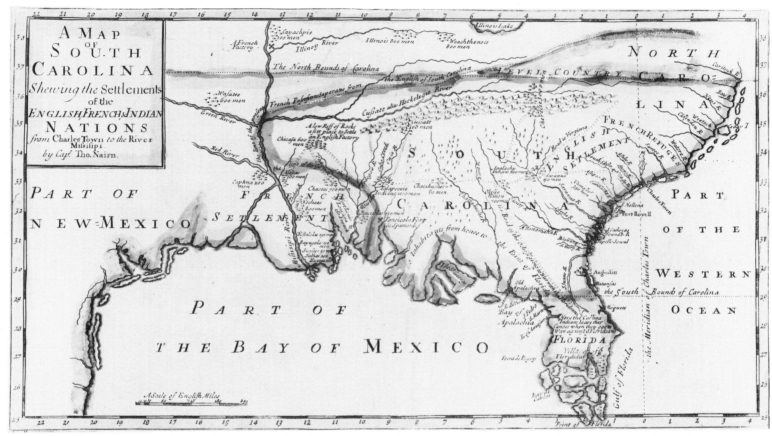

Figure 5 Thomas Nairne, *A Map of South Carolina* [detail] from *A Compleat Description of the Province of Carolina* (London, [1711]) [separately published]. Courtesy of the Library of Congress.

River; it was probably based on J. B. L. Franquelin's manuscript map of 1688, recording Louis Jolliet's explorations. Then in 1718 came a Delisle map that pushed back the boundaries of the British colonies still farther toward the Atlantic (fig. 6). In the South, he claimed that Carolina was named after Charles IX and settled by the French. There were encroachments in the North; Governor

William Burnet of New York wrote in anger to the Lords of Trade on 26 November 1720:

I observe in the last Mapps published at Paris with Privilege du Roy par M de Lisle in 1718 of Louisiana and part of Canada that they are making new encroachments . . . from what they pretended to in a former Mapp published by the same author in 1703

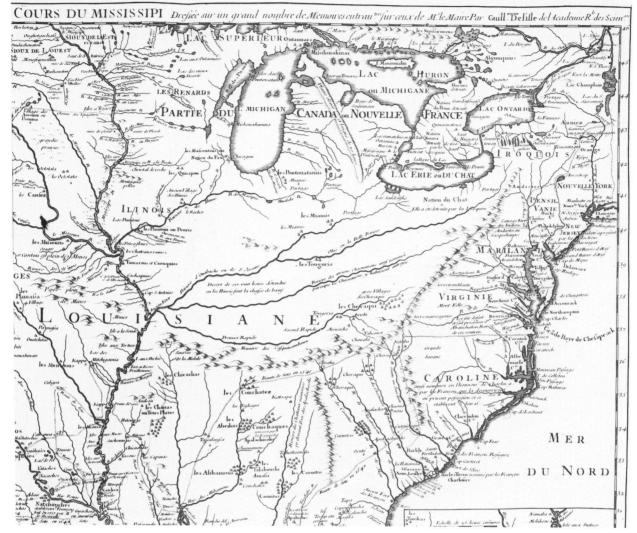

Figure 6 Guillaume Delisle [detail] from *Carte de la Louisiane* the Ayer Collection, the Newberry Library.
(1718), in Guillaume Delisle, *Atlas* (Paris, 1700–62). Courtesy of

9

of North America particularly all Carolina is . . . taken into the French Country and in words there said to belong to them and about 50 leagues all along the edge of Pensilvania & this Province . . . [9]

For many British mapmakers, Delisle's reputation and the geographical excellence of the map—J. G. Kohl called it "the mother and main source of all the later maps of the Mississippi"[10]—outweighed its territorial claims. A manuscript copy of it with an English translation, formerly owned by the Board of Trade and Plantations, is now in the Public Record Office. John Senex copied it, without acknowledging his source, in "A Map of Louisiana" in his *General Atlas* (London, 1721).

Herman Moll, a Dutchman in London as early as 1681, became the most vigorous cartographic antagonist of Delisle. His map of the British dominions in North America (1715) had already shown strong English counterclaims beyond the Appalachian mountain range to Lake Erie. He reproduced several sections of Crisp's map; he added to the Nairne inset the fighting strength of Indians of the region. On the detailed map of Carolina (bottom center) along the road leading north from Charles Town is "Phil, Pa." This apparently is a reference to a postal route by land to Philadelphia, for at the top right is a long legend on postal service in the colonies. This is one of the earliest references to post routes on an American map. One sentence reads: "The Western Post sets out from Philadelphia every Fryday leaving letters at Burlington and Perth Amboy and arrives at New York on Sunday night." (Apparently the mail between Philadelphia and New York is almost as rapid today as it was in 1715.) The view of Niagara Falls with beavers in the foreground at the right lower center is not original with Moll; he reversed an engraving on a 1713 map by the French royal geographer, Nicholas de Fer.[11] De Fer, in turn, had used the famous Hennepin picture (1697) of Niagara Falls,[12] adding the beavers. This is an instance of the connection between colonial-European trade and mapmaking. The enormous popularity of beaver hats in England had no small effect on the expansion of the American frontier, which followed the beavers as they retreated westward.

Five years later Moll published a map of North America (1720)[13] in which he specifically attacked Delisle and his 1718 map and called on English proprietors and merchants to note French "Incroachments" on British territory. New information on this map, he said in the cartouche, he acquired from "the original draughts of Mr. Blackmore, the Ingenious Mr. Berisford now residing in Carolina, Capt. Nairn and others never before publishe'd." It is an interesting study in map sources to see what these persons contributed. Lieutenant Nathaniel Blackmore of His Majesty's man-of-war *The Dragon*, referred to as surveyor general of Acadia, made a rough draft of the Nova Scotia coastline that he sent to the Board of Trade in 1715 (fig. 7).[14] Along the coasts on this chart appears a series of lines that may possibly be isobaths (depth-sounding lines), although they are inconsistent and not clearly related to the numbers that also appear on the chart. If the lines are isobaths, this is possibly the earliest use of this type of symbol in North American charting. Moll may have used this work in delineating Nova Scotia in both his 1715 and his 1720 maps; but he also used other sources or a different version of it, since details in the configuration of the peninsula differ from Blackmore's 1714–15 chart. Moll did use another more detailed survey by Blackmore in an inset, "The Harbour of Annapolis Royal," found both in his 1720 map and in his "A Description of the Bay of Fundy," which appeared in his *Atlas Minor;*[15] on the latter inset Moll adds "by Nathaniel Blackmore esqr." Many years later John Mitchell, in a note on his "A Map of the British and French Possessions . . . 1755," ridiculed the inaccuracies of Blackmore's "feigned Survey" of Nova Scotia and questioned the title of surveyor general.[16]

Moll's second source, "the Ingenious Mr. Berisford now residing in Carolina," is Richard Beresford, a wealthy Charles Town merchant and planter, who was one-time president of the commissioners of Indian trade, a member of the governor's council, and an

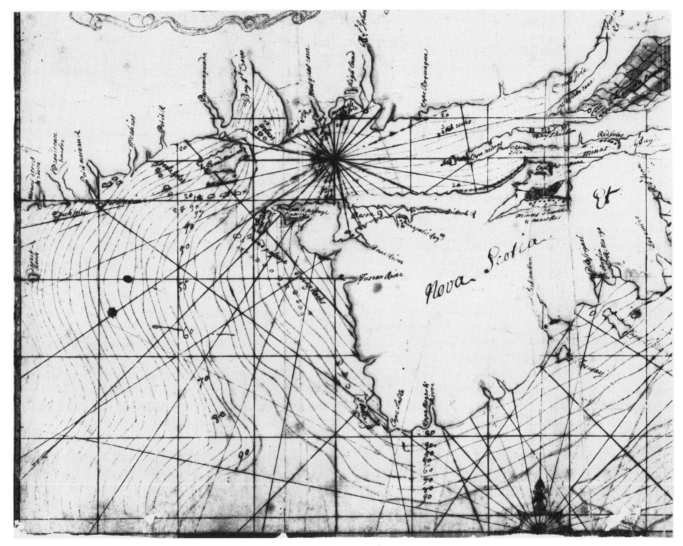

Figure 7 Nathaniel Blackmore [detail] from manuscript chart of Nova Scotia, etc. (1714–15) . Crown-copyright reserved. Courtesy of the Public Record Office, London (C.O. 700 N.S.4).

antiproprietary agent of South Carolina sent to England in 1714.[17] His father John had been surveyor general of the province. Beresford was therefore quite knowledgeable concerning affairs in Carolina, and Moll must have been in touch with him while he was in London in 1714 and 1715. The third source mentioned is Captain Nairne, whose 1711 inset of the Southeast is reproduced on Moll's 1715 "Dominions" map; on his 1720 map of North America he gives an entirely different conception of the Floridian peninsula and shows the route taken by Nairne, with the Yamassee Indians, Carolina allies, on a raid against the Indian tribes in Spanish territory. The map shows their canoe trip up the Saint John's, whence they portaged to the Kissimmee River, which flows into the Everglades. There they took thirty Indians captive for the slave trade. These maps of the British colonies by Herman Moll are some of the best of the period. His lettering is excellently clear, and he gives information of considerable historical value in his numerous legends.

On the 1720 map, between the Savannah and Altamaha rivers in Georgia, one can see Sir Robert Montgomery's abortive Margravate of Azilia.[18] In 1717 Sir Robert received this large grant of land from the Lords Proprietors. In his promotional literature he drew a map of the proposed colony, with Georgia marshes and pine forests laid out in neat squares (fig. 8). Each district of four hundred square miles is surrounded by a rampart with sixty-six fortified bastions in which the farmer-colonists live. Nothing so well illustrates his utter inability to understand the needs of an American frontier settlement in wild game country as that, among his divisions for livestock, one enclosure is set aside for a deer park. In the bottom left a hunter, presumably one of the landed gentry-to-be, is pictured shooting deer in his park. Fifteen years later Georgia was established under different auspices, but there is a startling similarity between Azilia and Peter Gordon's view of Savannah, the chief city, as it appeared in 1734 (fig. 9).[19]

Even more notable than Moll's maps, however, was a manuscript map presented to the Lords of Trade and Plantations about 1722 by the famous Indian fighter and expansionist, Colonel "Tusca-rora Jack" Barnwell. Historically, geographically, and politically, the Barnwell map is one of the most important maps of colonial southeastern North America.[20] Its numerous legends include information concerning Indian tribes, traders' paths, and settlements not elsewhere recorded except in its derivatives. The location of mountain ranges, the upper reaches of river systems, and the nature of land to the interior are here given in more accurate detail than in previous maps. The political plans of the South Carolina expansionists are indicated by the comments in the legends and the placement of forts. This manuscript, now in the Public Record Office, was used so heavily and constantly in the eighteenth century that it is almost illegible. Unfortunately, it was never published, and no other copy was known except in later derivatives, which usually omit the important legends.

In November 1969, Dr. Helen Wallis, the Superintendent of the British Museum Map Room, and I went to Chatsworth House, the estate of the Duke of Devonshire in Derbyshire, acting on a suggestion from Thomas Adams of the John Carter Brown Library. There we found a clear and beautifully preserved copy of this map, with the legends legible. It is signed "Hammercon fec. 1721," undoubtedly the copyist. A detailed study of the two maps shows it to be earlier than the Public Record Office copy. How, when, or why the 1721 Barnwell map got into the Chatsworth Library, T. S. Wragg, the Keeper, does not know.

In 1733 Henry Popple published a map of the British Empire in North America in twenty large sheets with a key sheet.[21] Popple used the Barnwell map, but not very intelligently or fully; the Lords of Trade and Plantations repudiated Popple's map, since some features seemed to militate against British territorial claims. In 1755 Dr. John Mitchell's map of the British and French Dominions in North America used many of Barnwell's legends and geographical details.

Since the time of Jacques Cartier explorers have referred to maps drawn by Indians. Baron Lahontan, an early eighteenth-century mapmaker and observer of the Indians, said that "they draw the most exact Maps imaginable of the Countries they're

Figure 8 Robert Montgomery [plan of "Azilia"], in: Sir Robert Montgomery, *A discourse concerning . . . a new colony* (London, 1717). Courtesy of the Ayer Collection, the Newberry Library.

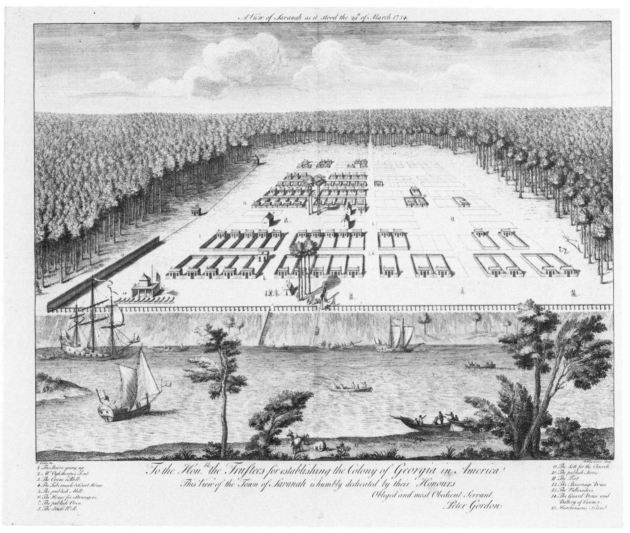

Figure 9 Peter Gordon, *A View of Savanah . . .* (1734) [separately published]. Courtesy of the John Carter Brown Library.

acquainted with, for there's nothing wanting in them but the Longitude and Latitude of Places: . . . These *Chorographical Maps* are drawn upon the Rind of your *Birch Tree*; and when the Old Men hold a Council about War or Hunting, they're always sure to consult them."[22] Later in the century, Governor Thomas Pownall wrote that "The Habit of travelling mark to him [the Indian] the Distances, and he will express accurately from these distinct Impressions, by drawing on the Sand a Map which would shame many a Thing called a Survey. When I have been among them at Albany and enquiring of them about the Country I have sat and seen them draw such."[23] Very few examples of Indian birchbark and deerskin maps have survived. The examples now in the Public Record Office, originally drawn on deerskins by an Indian chief for Governor William Lyttleton of South Carolina about 1724, exist now only as tracings on paper. This disappointment was assuaged, however, by an Indian birchbark map that Dr. Wallis and I came across in a collection since acquired by the British Museum. The bark map was found in 1841 by Captain Bainbridge of the Royal Engineers on "the ridge" somewhere between the Ottawa River and Lake Huron.[24] It shows part of the route taken by the Indians along the waterways between Lake Huron and the Saint Lawrence River. There is a note by Captain Bainbridge accompanying the map that transcribes the route and treats the map as an example to young engineers in their early efforts at surveying. Maps drawn on bark, chiefly on birchbark, were common among the Indians in the North, who are known to have taken whole rolls of maps on their wanderings.

CAROLINA MAPS OF THE SECOND HALF OF THE EIGHTEENTH CENTURY

The third quarter of the eighteenth century saw rapid advancement in surveying methods and in the mapping of the Southeast. The most prolific mapmaker in the Southeast at this time was William Gerard de Brahm, a German Protestant who had been a Captain of Engineers under Emperor Charles VI. With De Brahm, we turn from the amateur to the professional, from general outlines of a region to topographical accuracy. His "Map of South Carolina and Part of Georgia" (1757,)[25] in its depiction of the size and shape of the coastal islands and tidal lagoons and its direction of the rivers and location of settlements, is far superior to any mapping of the southern district that had appeared before. An interesting detail is the analysis of the soil along the South Carolina-North Carolina boundary, an early example of its type (fig. 10). In 1764 De Brahm was appointed Surveyor General of the Southern District, the region south of the Potomac and from its headwaters west to the Mississippi. During the next six years, in addition to other activities, he made coastal surveys from South Carolina southward and around the Florida peninsula that were later incorporated in charts of *The Atlantic Neptune.* In 1772 appeared his *Atlantic Pilot,* published in London, avowedly a manual of sailing instructions but also a discussion of the supposed course and origin of the Gulf Stream. It is a pioneer appraisal of this important subject. During 1771–73 he composed a remarkable "Report of the General Survey in the Southern District," part of which was published over seventy-five years later (1856) as De Brahm's "Philosophico-Historico-Hydrogeography of South Carolina." Plowden Weston, the editor, who mentions De Brahm's death in 1799 in Philadelphia, wrote that he "lived in the memory of persons now alive, much addicted to alchemy and wearing a long beard."[26]

In 1770 Captain John Collet, commandant of Fort Johnston on the Cape Fear River, published in London "A Compleat Map of North Carolina from an Actual Survey" (fig. 11). The northern half of Collet's map is based on almost twenty years' work by William Churton, surveyor of the Granville District, who had many years earlier supplied the information for the part of North Carolina included in Joshua Fry and Peter Jefferson's "Virginia" (1751).[27] Of the manuscript left by Churton at his death in 1767, Governor William Tryon of North Carolina said, ". . . I am inclined to think there is not so perfect a draft of so extensive an interior country in any other colony in America."[28] Captain Col-

Figure 10   W. G. De Brahm [detail] from *A Map of South Caro-lina and a Part of Georgia* (1757), in: Thomas Jefferys, *A General Topography of North America* (London, 1768). Courtesy of the Ayer Collection, the Newberry Library.

let's contribution to the map was primarily in the southern and coastal area. It was the basis for later maps of the province and state for nearly forty years.

Claude Joseph Sauthier was a French surveyor and draftsman and another protégé of Governor Tryon, for whom he made a series of plans of ten towns in North Carolina, such as those of Hillsborough, later to be the temporary capital of North Carolina, and Edenton on Albemarle Sound. No other such series of careful plans of small towns exists, as far as I know, before the Revolution. The remarkable work of Sauthier, hitherto relatively unknown, spans the later colonial period (see pp. 72–74).

VIRGINIA-MARYLAND-DELAWARE

Throughout most of the eighteenth century, the Virginia-Maryland-Delaware region formed a geographical and cultural if not a political entity. Most of the maps published emphasized Chesapeake Bay and the adjacent tidewater lands. Until the 1670s, all European mapmakers followed the 1612 map of Virginia that Captain John Smith had made under extraordinarily difficult conditions.[29] During the summer of 1608 he navigated the waters of the Chesapeake in a primitive barge, attacked by Indians, poisoned by a sting ray, and hampered by the sickness of his small crew. The result of his three months' work, miscalled a canoe survey, was a good map of the Chesapeake and many of its tributary rivers.

In 1653 Augustine Herrman, a Bohemian who had arrived in New Amsterdam about six years before, offered the Dutch governor, Stuyvesant, a plan to map the head of Chesapeake Bay. Stuyvesant was not interested; but the governor of Maryland was, and offered him a manor, Bohemia Manor on Bohemia River, for the work in constructing the map. Herrman finished a preliminary map, apparently of Maryland only, by January 1661. Continuing, he undertook extensive travels and surveying and had gathered all the information for compiling his enlarged map by 1670. "Virginia and Maryland As it is Planted and Inhabited this present Year 1670" appeared in London, engraved by W. Faithorne and pub-

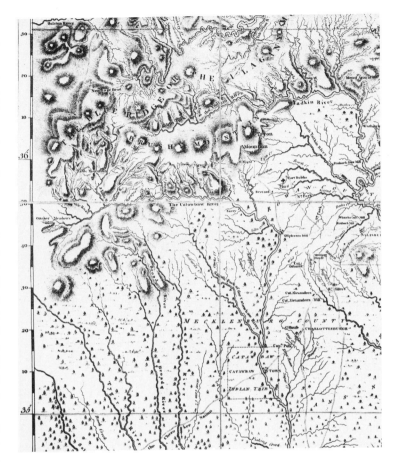

Figure 11 John Collet [detail] from *A Compleat Map of North Carolina . . .* (London, 1770) [separately published]. Courtesy of the Ayer Collection, the Newberry Library.

lished in 1673 by John Seller, Hydrographer to Charles II.[30] Carefully prepared and detailed in its nomenclature of plantation owners and of counties, Herrman's map was the source of subsequent maps of the area for eighty years, except for hydrographic surveys of the bay.

As the middle of the eighteenth century approached, the westward push of settlement toward the Valley of Virginia and the urgent need of better information concerning western routes by the colonial government and by the Board of Trade and Plantations in London resulted in the notable work of Joshua Fry and Peter Jefferson, "A Map of the Inhabited part of Virginia containing the whole Province of Maryland, Part of Pensilvania, New Jersey, and North Carolina . . . 1751."[31] Joshua Fry, professor of mathematics at the College of William and Mary, and Peter Jefferson, father of a more famous son, continued with corrections and additions to their map. These they sent to London together with Captain John Dalrymple's table of distances between towns in Virginia and in other colonies. This state of the map was published in 1755; it was brought out again in 1775 with no significant change on the plates but the new date.

John Henry's "Virginia" (1770), based on additional material, was not received with like enthusiasm.[32] A contemporary, Governor Thomas Pownall, condemned it as "a very inaccurate Compilation; defective in Topography: and not very attentive even to Geography," and subsequent writers have frequently noted its errors in the location of towns and courthouses. The new information that it gives, however, includes the location of many settlers and plantations by name, the boundary lines of Virginia counties, and identification of areas west of the Blue Ridge. Its richness of detail makes John Henry's map of considerable historical if not of geographical value.

During the eighteenth century, no important maps of Maryland and Delaware appeared, except those including them with Virginia. Although the division between the Northern and Southern Departments of the Indian Superintendents and of the Surveyors General was defined by the British government as the Potomac and from its headwaters westward to the Mississippi River, Maryland was closer to Virginia than to Pennsylvania geographically,

culturally, and commercially. Chesapeake Bay was a dominant factor in the life of both.

The boundary lines between Maryland, Delaware, and Pennsylvania were a constant source of contention. The surveys that resulted were recorded on maps from Taylor and Pierson's "circular" boundary line map of 1701, marking the twelve-mile radius from Newcastle that still remains the northern boundary of Delaware, to the Mason and Dixon survey of 1767, which extended from Cape Henlopen to the crossing on Dunkard Creek in the Alleghenies. The Mason-Dixon line became the very symbol of the division between the North and the South.[33]

LATER REGIONAL MAPS OF THE SOUTHEAST

In 1763 the Treaty of Paris gave England practically all North America east of the Mississippi as well as Canada. It was of the utmost importance to the authorities in London to have some geographical knowledge of this terrain and of the Indian tribes that they wished to pacify. Lord Hillsborough sent urgent instructions to John Stuart, Superintendent of Indian Affairs in the Southern Department, for "an accurate general map" on which the famous Proclamation Line of 1763, which divided the colonial settlements from the Indian territories west of the Appalachian watershed, was to be drawn. No such map existed, and the hastily conceived Proclamation Line between the English settlements and Indian territory, based upon inadequate geographical information and lack of understanding of current frontier conditions, became untenable. It was Stuart's great achievement to establish in twelve years a workable Indian boundary line from the Ohio River to the Gulf by treaties with the Indians and to construct an excellent map of the entire southeastern region. John Stuart, a Scot born in 1718, was an Indian trader in Charles Town, South Carolina, during the 1750s; he was an able and expertly informed Superintendent of Indian Affairs from 1763 until his death in 1779.[34] A mapmaker himself, he had a full realization of the importance of correct maps as an adjunct in solving the problems of his administration. Fortunately, he had for the coastal region the established work of Joshua Fry, Peter Jefferson, John Collet, William De Brahm, and James Cook;[35] with their maps he collated the surveys of some younger men, such as Bernard Romans[36] for Florida, Henry Yonge for Georgia, and David Taitt, John Donelson, and Elias Durnford for the Gulf Coast and the rivers flowing into it. With the aid of his department's most talented cartographer, Joseph Purcell, he produced a series of large wall maps, about six feet by six feet, that provided a wealth of information about the roads, trails, mountain ranges, British settlements, colonial boundaries, and the location of Indian villages and tribes. The first of these he completed in 1775 and sent to London early in 1776. These maps are known as the Stuart-Purcell maps. One of the finest and best preserved of them is in the Newberry Library; it is one of the glories of the Ayer Collection (fig. 12).[37]

One legend of the Stuart-Purcell map is of particular interest since it illustrates the use of old maps as legal evidence: it is the statement concerning part of the Indian boundary line signed at the Treaty of Picolata in 1765; "The Indian Boundary is by Treaty the West Bank of the St. Johns to its Source. from there Southward and round the Cape, is Regulated by the Flowing of the Tide." This means that the Indians at Picolata granted the British by treaty all of the land along the Florida coast up to the point on each river to which the tide rose, but reserved the interior for themselves. A few years ago I received a letter from the Assistant Attorney General of the United States. It said, in summary: The Seminole Indians have sued the United States for the State of Florida. Treaties made by the Indians before the acquisition of territory by the United States are valid. The Indians cannot legally lay claim to land ceded to the Spanish and British by such treaties. Will you be willing to find these treaties and accompanying maps, draw the boundaries on a large-scale map, and defend them point by point before the Indian Claims Commission? Two years later, during which period I was gathering material, documenting the extent of tidal flow up every river around the penin-

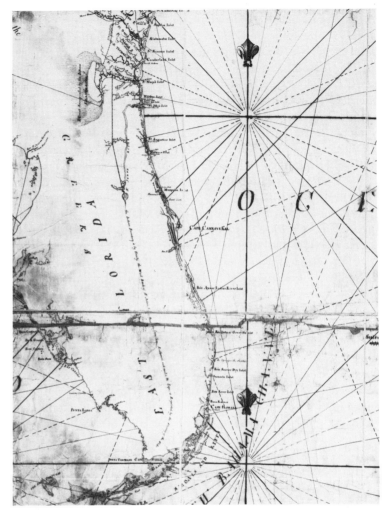

Figure 12 John Stuart and Joseph Purcell [detail] from *A Map of the Southern Indian District of North America* . . . (1775) [manuscript]. Courtesy of the Ayer Collection, the Newberry Library.

sula, and preparing the maps to present in court, the case was tried by the commission and later by the United States Court of Claims.[38] There were several treaties involved, some of which were disallowed because the Seminoles were not signatories. They were awarded forty million dollars; at least they will receive what remains above their lawyers' fees after Congress votes to appropriate the money.

Joseph Purcell arrived in East Florida as a youth with his family in a large group of Mediterranean peoples brought over by Dr. Andrew Turnbull to found the new colony of Smyrna. Bernard Romans referred to him as "an excellent young man, a Minorquin."[39] He was employed by De Brahm, from whom he probably learned his cartographic and surveying skills; later he became surveyor of the Indian Department. In a letter to Dartmouth dated 23 October 1773, De Brahm complained that his "geometer" Joseph Purcell had secretly taken copies of his surveys and joined the staff of John Stuart, as had "one Bernard Romans."[40] After Stuart's death in 1779, Purcell continued his position as Surveyor of the Indian Department under Colonel Thomas Brown, who succeeded Stuart as Superintendent of Indian Affairs. In 1781 he completed the last and most detailed of his great series, the Brown-Purcell wall map of the Southeast, now in the Public Record Office.[41]

Purcell is one of the British colonial cartographers who continued to practice his trade after the Revolution. After the British evacuated Charleston at the end of the war, he was seized, imprisoned, and later paroled. He became a citizen and continued to live in Charleston.[42] In 1788 he made "A Map of the States of Virginia, North Carolina, South Carolina and Georgia . . . By

Joseph Purcell . . . New Haven, 1788," which was printed in Jedediah Morse's *American Geography* in 1789 and again (with revisions in a new edition of the map) in 1792. Morse's preface refers to the map as "compiled from original and authentic documents, by Mr. Joseph Purcell of Charleston, South Carolina, a Gentleman fully equal to the undertaking, and is the most accurate yet published respecting that country, on so small a scale."[43]

There are many maps and plats of small land holdings made by Purcell during the 1780s and 1790s in the collections of the South Carolina Historical Society in Charleston and elsewhere. One is a beautifully drawn manuscript plan in the Public Record Office of an estate on the Saint John's River at Beauclerk's Bluff.[44] Made in 1771 while he was still in De Brahm's employ, the map shows he was already a skilled craftsman in his trade. The Saint Phillips Register in Charleston records his marriage to Mrs. Ann Bonsall on 20 September 1792.[45] On 13 June 1797 he signed a five-page document, now in the South Carolina Historical Society, concerning the western territory of Georgia, in which he gives his position as deputy surveyor for the State of South Carolina. He continued to serve as surveyor into the nineteenth century.

From the First Lords Proprietors' map of Carolina of 1672, with its geographical misconceptions of a desert, swampy savanna, and great lake filling the Southeast, to Purcell's maps, with their wealth of historical, cartographical, ethnological, and archeological information, is a tremendous advance. By the Revolution, not all of the Southeast had been surveyed; but its boundaries, coasts, rivers, mountain ranges, and settlements were known and had been mapped by British cartographers.

Mapping the Northern British Colonies

*Expanding Knowledge of the Interior and
the Settlement of the North*

The history of British mapmaking in the northern colonies is more complex than that of the southern colonies for several reasons. One is the diversity of the region. This chapter treats of British mapmaking in the Middle colonies (Pennsylvania, New Jersey, and New York), New England, Canada, and the trans-Allegheny region, although the last two areas were primarily in the hands of the French until after the middle of the century. Obviously the North was colonized by a greater diversity of powers than the South. Cartographically, from earliest times, the South has been treated far more as a unit.

Another point of contrast between the British cartography of the North and South arises from their differences of cultural pattern. In the South the plantations were prominent on the maps; the subscribers to a map were often the landowners themselves. In the North small towns and settlements were the more prominent features; in New England the churches or meetinghouses are often indicated. Boundary disputes play a part in the mapping of both regions.

Except for Pennsylvania and New York City, there are comparatively few cartographic studies for the northern colonies. A great need exists, for example, for someone to make an analytical list of the maps of New England, and there are many other desiderata.

In the South the course of cartography moved from the provincial to the regional map that culminated in the great series of Stuart, the Indian Superintendent, and his assistant Purcell. In the North a remarkable group of independently sponsored regional maps appeared around 1755, such as the Evans "Middle British Colonies" and the Douglass "New England," whereas the best maps of individual colonies came after 1755 but were not incorporated into new and better regional maps.

THE MIDDLE COLONIES: NEW JERSEY, PENNSYLVANIA, NEW YORK

The Middle Colonies form a kind of mapping unity, although Pennsylvania and New York each had a prolific cartography of its own.

The first map of New Jersey under that name was produced by John Seller and William Fisher between 1664 and 1674.[1] In 1664 Charles II granted to his brother James, the Duke of York, the area between the Connecticut River and the Delaware. This was five months before the English capture of New Amsterdam; James, in turn, two months before the Dutch capitulation, granted the land between the Hudson River and the Delaware to two adherents, Sir John Berkeley and Sir George Carteret. He named the tract New Jersey in recognition of Carteret's lieutenant governorship of the island of Jersey during the Commonwealth. The Dutch unexpectedly recaptured New Amsterdam in 1673, but the English regained it by treaty in the following year. The Seller map was made as part of a vigorous promotion campaign by the Quakers. It is a large and handsome map that was revised several times; in 1677, in the fourth state, side plates enlarged the map, which also had a promotion tract pasted on the bottom.[2] The geography shows both lack of knowledge of coastal configuration and imaginary additions, such as a nonexistent lake flowing into a branch of the Delaware and the union of the Hudson and the Delaware rivers at their headwaters. The early states had a view of New Amsterdam as it was about 1650, taken from a Dutch map by

Visscher; this was changed later to Allardt's "Restitutio" view of the city of about 1673.[3] The Seller map itself was evidently taken directly from a Dutch map; it is beautifully designed in the skillful manner of the golden age of Dutch cartography, with details of animals of the region, deer hunting, and palisaded Indian settlements.

For nearly a hundred years after Seller's map, no good separate map of New Jersey based on surveys appeared, although in 1755 both John Mitchell's "British and French Dominions" and Lewis Evans's "Middle British Colonies" included New Jersey in the best delineation of the province published up to that time.[4] In 1768 Jefferys, the London atlas maker, published "The Provinces of New York and New Jersey . . . drawn by Capt. Holland." But Holland, the Surveyor General of the Northern District, vigorously denied that he had anything to do with it.[5] Actually this map is little more than an enlarged (or blown-up) copy of New Jersey in Lewis Evans's 1755 map, with the incorporation of some surveys made by Holland in the Province of New York.

By contrast, William Faden's "The Province of New Jersey, Divided into East and West, commonly called The Jerseys" (1777) demonstrates well the notable advances both in surveying and in knowledge of the country since the Evans map.[6] This is a finely engraved work, of great detail, with hachuring for hills and ridges. It was based on a 1769 survey by Lieutenant Bernard Ratzer of the 60th American Regiment, who had been commissioned to survey the disputed partition line between New Jersey and New York; the northern part of the map was based on a survey made by Gerard Bancker. The longitudes had been checked by astronomical observations, the surveys show careful and exact professional work, and the map would not be far out of place in a modern atlas.

From its beginning, the Province of Pennsylvania fared better cartographically than New Jersey.[7] In April 1682, William Penn appointed Thomas Holme as surveyor general to plan the colony and a chief city. In 1683 appeared Holme's "A Portraiture of the City of Philadelphia," a plan for the city and for allotting land. In 1687 came Holme's "Map of the Province of Pennsylvania," covering an area extending thirty miles west of Philadelphia, naming the counties and townships, and labeling the plantations and farms with the names of the owners, as Maurice Mathews had done in South Carolina.

Postponing until later a discussion of Evans's maps, which cover a larger area than Pennsylvania, one comes to the work of Nicholas Scull, "Map of the Improved Part of Pennsylvania," published in 1759. Scull had been surveyor general of Pennsylvania since 1748, and with G. Heap had published "A Map of Philadelphia, and Parts adjacent" about 1750. Active in the councils of the provincial government, especially in its relationship with the Indians, he knew and had surveyed the frontiers of the province. He and Evans had aided and exchanged information with each other; Scull was a better surveyor than Evans and his map possessed unusual accuracy for that time. Its considerable detail was engraved with sharpness and clarity. Scull's grandson, William Scull, produced an even better map, based on Nicholas's, eleven years later, in 1770. It extended the surveying of the western frontiers, showed the final line run in 1763–67 by Mason and Dixon between Maryland and Pennsylvania, added roads, trails, and place-names, and acknowledged the aid received from the plans and information of numerous persons.

New York became a province of the British Empire in America in 1674, but the Dutch had already charted its spacious harbor and drawn many maps, often more artistic than accurate, of the country around the harbor and up the Hudson. It remained for the English, however, to explore thoroughly and to map the interior. The cartography around New York City, both before and after the English occupation, has been carefully examined: I. N. P. Stokes's monumental six-volume *Iconography of Manhattan Island* has covered its history, pictorial delineation, and mapping with a thoroughness no other area in this country has received. Yet even the *Iconography* does not exhaust the record, for I do not find mentioned the maps of Colonel Römer, the first cartographic records of the area in the eighteenth century.

William Wolfgang Römer, a colonel of engineers in the British army, was sent to New York in 1698 to oversee and strengthen the fortifications of the city and province. He remained in New York for five years. In 1700 he made several maps, including one of the entrance to New York Harbor. The next year he submitted to Governor Bellomont his recommendation for building forts at the Narrows, Sandy Hook, and elsewhere. He also made a journey to the interior of the province, to examine places for defense against the Indians and the French. In his maps of this expedition, which show the influence of previous Dutch mapmakers, he gives details of his route up the Hudson and Mohawk rivers toward the Ontario country, with a long explanatory legend in unorthodox spelling (fig. 13). In Boston, Römer supervised the construction of Fort William on Castle Island, completed by 1703. In 1705, he "made and survaied" a large-scale "Icqnographical Draft [ground plan] of Castle Island" in Boston Bay, with profiles of Castle William.[8]

Meanwhile, in England, commercial publishers were making greatly improved maps from the drafts of the new surveys as they arrived. Thomas Jefferys, geographer to the king, was especially active. He published John Montresor's "New York and its environs" in 1766. It is the best work on its subject up to that time, although Montresor, a British engineer, was hampered in his surveys of New York City by the danger of physical violence from the patriots enraged by the Stamp Act. He found thirty-one errors on the first impression that was brought to him from the engraver. Lieutenant Bernard Ratzer made a survey of New York City in 1766–67 that resulted in another fine map of New York published by William Faden, successor to Thomas Jefferys, in 1776, with a beautiful view of the city and harbor (fig. 14).[9]

Around 1773 Claude Joseph Sauthier, Governor Tryon's surveyor, was at work on a large-scale map entitled "A Chorographical Map of the Province of New York," published by William Faden in 1779.[10] This includes present-day Vermont, at that time part of a disputed area between New York and New Hampshire; both Tryon and Sauthier had large landholdings there. Sauthier's chorographic map is the culminating cartographic achievement of maps of New York, rapidly improving in quality during the 1770s.

It shows the clarity and delicate skill of Sauthier's work as a draftsman.

In 1755 appeared an important map that covered the entire area from the Connecticut River and Lake Ontario in the north to Carolina in the south. This was Lewis Evans's "A General Map of the Middle British Colonies in America," the culmination of a series of maps produced by him. He had published "A map of Pensilvania, New Jersey, New York . . ." in 1749, so filled with explanatory and informative legends scattered across its face that, valuable as his comments are, the total effect is one of confusion. He is the first to note, for example, that storms along the coast travel in a northeast and easterly direction, "a day sooner in Virginia than in Boston," a phenomenon we see now portrayed on our daily television weather forecasts. Evans's knowledge and ability were recognized by the General Assembly of Pennsylvania, which commissioned him to do a map of the province. A preliminary draft, made in 1751, was sent to the Board of Trade in London; it was used as the basis of instructions for General Edward Braddock. Evans continued to gather additional information for several more years. He surveyed the settled part of the province with the aid of the surveyor Nicholas Scull. He took the latitude of many places and computed the longitude with the aid of calculations of observers in Philadelphia and Boston. Evans learned about the western parts of the province to the north and west of the Ohio Valley from the explorations of the famous scout Christopher Gist, the land-promoter Dr. Thomas Walker, and many others. He published his map, with a valuable accompanying essay, in 1755.[11]

Evans's interest ran far beyond topographical details. He used his map to show the importance of the Ohio Valley in the struggle for supremacy against the French, and was a conservationist before his time in his comments on such matters as erosion of the soil. His tragedy was that, being too outspoken, he antagonized vested and imperialist interests by his limitations of the boundaries of Pennsylvania and by conceding the claims of France to the land north of Lake Ontario. In addition, he angered Governor Robert Morris of Pennsylvania, who imprisoned him on a false charge of libel.

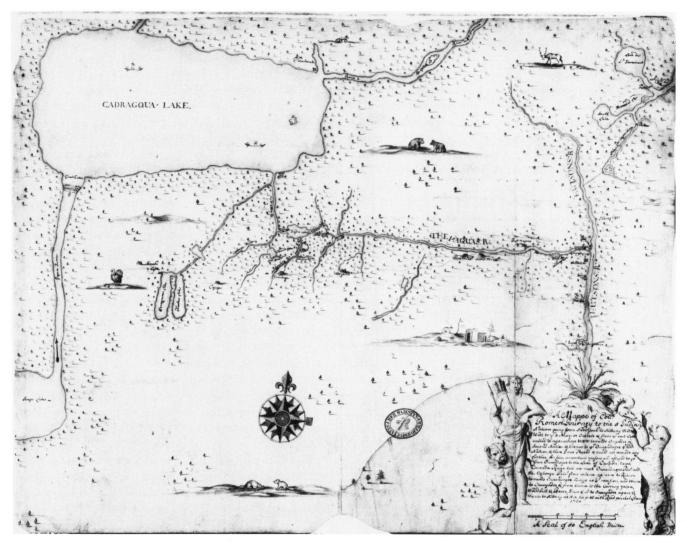

Figure 13 W. W. Römer, *A Mappe of Coll.* [*sic*] *Romer his Journey to the Indian Nations . . .* (1700) [manuscript]. Crown-copyright reserved. Courtesy of the Public Record Office, London (C.O. 700/N.Y. 13A).

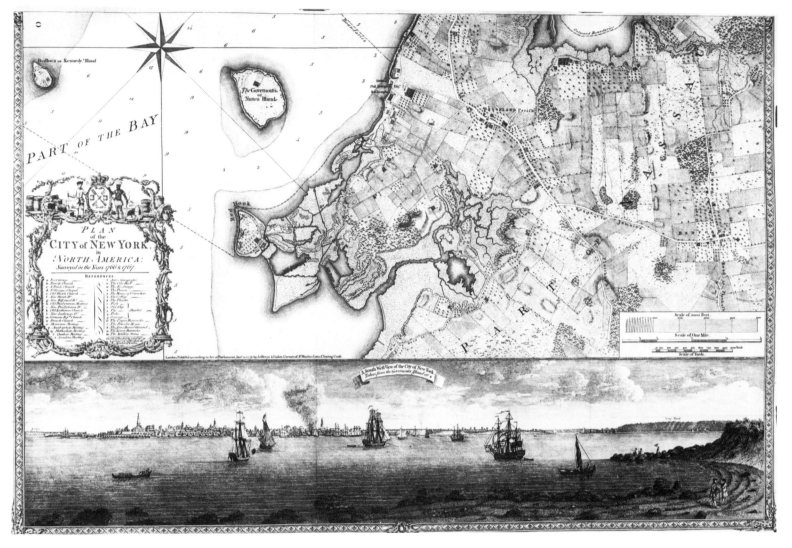

Figure 14 Bernard Ratzer [detail] from . . . *Plan of the City of New York . . .* (1776), in: William Faden, *North American Atlas* (London, 1777). Courtesy of the Ayer Collection, the Newberry Library.

On his death in 1756 he left his family in poverty. His map remained his greatest achievement. Unfortunately, it was so good that pirated editions began to appear shortly after its publication. In 1776 Thomas Pownall, former governor of Massachusetts and a strong friend of Evans, made a revised edition with numerous improvements, pledging any profits to the daughter of Evans.

NEW ENGLAND

Explorers had sailed along the New England coast and put it on maps from the earliest times; the interior, however, was neither known nor mapped as early as Canada or the South. In the eighteenth century, settlements had spread rapidly, although without the surveys during the first half of the century that one might expect. Connecticut, with its complicated boundary disputes, produced two maps of interest. Before surrendering their old patent to the Crown, the Council of the Plymouth Company in England determined to divide the seacoast of southern New England among themselves. Some of the territory was granted twice and even thrice over, which later resulted in prolonged legal disputes. The land between the Connecticut River and the Narragansett River, for sixty miles north from the seacoast, was granted to the Marquis of Hamilton. It was almost sixty years later that the last of a series of petitions by his descendants was settled claiming the so-called county of New Cambridge. In 1697 the Duchess of Hamilton accompanied her claim with a beautiful map (fig. 15), but the Lords of Trade decided the case against her.

No sooner was this matter disposed of than an Indian claim to some of the same land arose to plague the landholders of Connecticut. The Mohegan Sachem Uncas (not the last of the Mohicans) claimed that his tribe should be compensated for its hereditary lands, which covered part of eastern Connecticut, including much of the area between the Connecticut and Quinebaug rivers. His case was upheld by Governor Dudley's court in 1705, although the matter was not finally adjudicated until 1771. A survey of the Mohegan boundaries used in the claim, now in the Public Record Office, gives topographical details and the names of early settlements such as York, Lebanon, Colchester, Preston, Canterbury, and Dudley found on no earlier map that I have seen. John Chandler, who made the map in 1705, was town surveyor of Woodstock, Connecticut. Six years later he became a judge of the General Court in Boston. In the northeast corner of the map lies "Man hum squag," a lake that later developed the name of Chargoggagoggmanchaugagogcchaubunagungamaug.[12] In 1737 Gardner and Kelloch (Kellogg) made an extensive survey of Connecticut of which no extant copy is known; but their surveys were used in the important maps of New England by Douglass (1753) and Jefferys-Green (1755).[13] Thomas Kitchin engraved a map of Connecticut and Rhode Island for the *London Magazine* in 1758 that was based on the Jefferys-Green "New England" but added county and township lines.[14] This was the earliest map printed primarily of these two colonies (which were often thus combined). It was followed by Moses Park's "Connecticut" (1766), engraved by Abell Buell; the first map designed, engraved, and published in Connecticut, it was made at the request of the Earl of Halifax and for the postal service.[15]

In Rhode Island, two maps in particular provide a good example of the contrast between the mapmaking in the early and late years of the century (figs. 16–17). The first, surveyed in 1720 by John Mumford, was approved by the General Assembly and signed by Governor Samuel Cranston.[16] The second is a chart of Narragansett Bay by Charles Blaskowitz, published in Faden's *North American Atlas* in 1777. Mumford's map is little more than a careful sketch with a few details added to the surveyor's lines and the outlines of the islands and the coast. Blaskowitz, who was one of the ablest military surveyors for the British army during the Revolution, records his careful work on the islands in the bay, notes the intricacies of bays, coves, and creek mouths, and gives the soundings for the main channels and hachuring for the elevations and bluffs. The map has a large cartouche and extensive annotations; the draftsmanship is superior, and the engraving is executed by a professional in William Faden's London shop.

While Massachusetts is not represented on separate maps in the

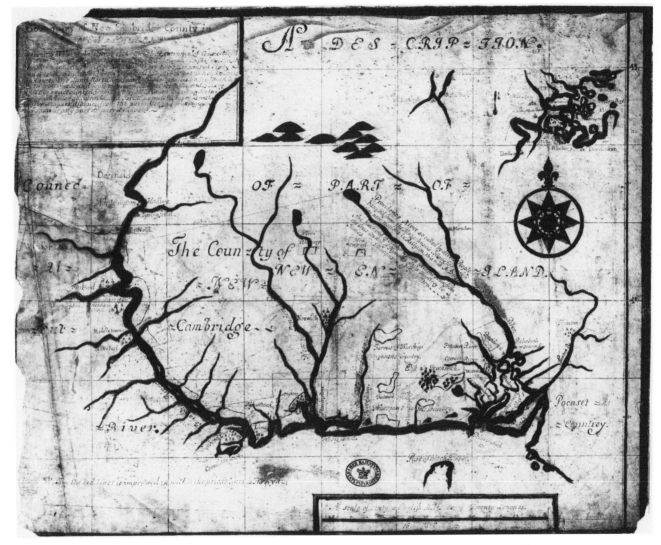

Figure 15   Boundings of New Cambridge County in New England (1697) [manuscript]. Crown-copyright reserved. Courtesy of the Public Record Office, London, (C.O. 700/Connecticut).

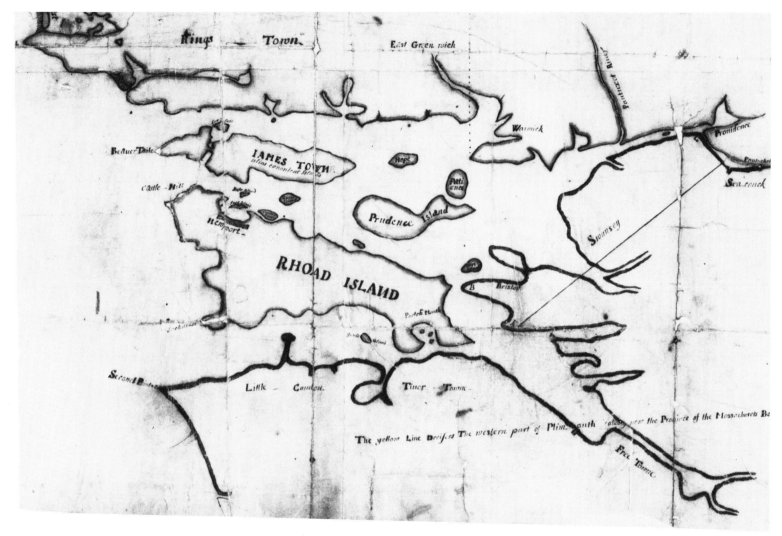

Figure 16 John Mumford [detail] from *Map of Rhode Island* (1720) [manuscript]. Crown-copyright reserved. Courtesy of the Public Record Office, London (C.O. 100/Rhode Island 1).

colonial period, maps of Boston and of the vicinity around it rival in number those of New York and Philadelphia, and surveys of boundary lines and of small areas are not lacking. The reason for the absence of maps of Massachusetts alone is probably simple: to the west it extends within fifteen miles of the Hudson River; to the south, Martha's Vineyard lies below the latitude of most of the Connecticut coastline; Cape Cod projects into the Atlantic east of the other colonies; and to the north and northeast Massachusetts claimed Maine as under its jurisdiction. A sheet that shows all of Massachusetts also includes the area of the other New England colonies. Massachusetts cartography will be treated later under general maps of the region; here I am going to discuss a special group of local maps made in the 1760s.

This collection of maps of Massachusetts and its province, Maine, belonging to Sir Francis Bernard, the royal governor of Massachusetts Bay (1760–69), is now preserved at Nether Winchendon House near Aylesbury, Buckinghamshire. A preliminary catalogue appears as Appendix A to this volume. Edward Miller, Assistant Keeper of the British Museum State Paper Office, mentioned to me some American maps he had seen there in the home of Dr. Spencer Bernard, a descendant of Sir Francis. In November 1969 I wrote to Dr. Bernard, asking if Mrs. Cumming and I might bring along Dr. Helen Wallis, Superintendent of the British Museum Map Room, to examine his maps. We reached Nether Winchendon House, after a ride through the beautiful Chilterns with their brick, half-timbered villages, and found it to be a charming small manor house, set among fields and walled gardens. In the twelfth century it had been a small monastery, then a Tudor dwelling, to which had been added eighteenth-century enlargements in the Strawberry Hill Gothic style of the period. Here, in 1771, Governor Sir Francis had retired after his return in 1769 from America. He spent his last years in Aylesbury. Dr. Spencer Bernard, a distinguished physician who had been pathologist to Buckingham Palace, brought out for us a collection of maps that, in purpose and type, differed so markedly from the more usual military, coastal, and general colonial maps of the time that it

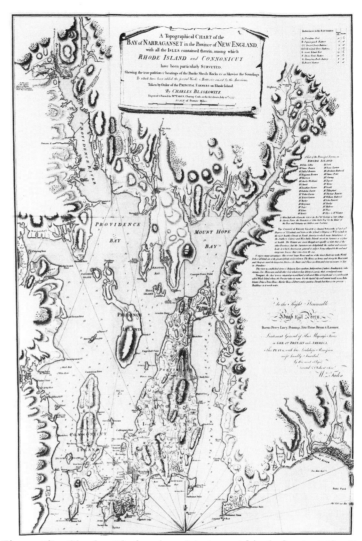

Figure 17 Charles Blaskowitz, *A topographical chart of the Bay of Narraganset . . .* (1777), in: William Faden, *North American Atlas* (London, 1777). Courtesy of the General Collection, the Newberry Library.

stands out in both interest and importance. These were domestic maps, of a gentleman's estates and the roads to them.

Sir Francis Bernard had been appointed governor of Massachusetts Bay in 1760 after a successful two-year governorship of New Jersey, during which he had made a workable treaty with the Mohawk Indians. His initial relationship with the people and the Assembly of Massachusetts was friendly; the assembly thought so well of him that two years after his coming they voted to award him "the Island of mount desert lying north eastward of penobscot bay . . ." in Maine. He was not a wealthy man; he had ten children, and was most anxious to be able to leave them a landed estate. He was also sincerely interested, as he said in a letter to Lord Barrington, in helping "to people one spot of the Vast wast of his Majesty's Dominions."[17] He had the whole region surveyed and mapped, employing the professor of mathematics at Harvard College.[18] Governor Pownall, in making his careful composite map of New England and the Middle Atlantic colonies in 1776, says of this part of the Maine shore, "The Coast included within these marks [two asterisks] is copied from Governor Bernard's Surveys, including Mo. Desart Id. &c."[19] Evidently he thought them accurate. Governor Bernard laid out streets on his island, and promised to install there, in the summer of 1764, an immigrant German congregation of eighty families. He took a leading role in the Kennebeck Company, which had marked out twelve townships "upon the continent adjoining to the Island." He divided lots into narrow waterfront strips along the shore and the banks of the Penobscot River that would have done credit to a Miami realtor. Alas, his enthusiasm made him disregard the fact that the Lords of Trade and Plantations had never confirmed Masschusetts Bay's right to a title to this territory. This title was never granted. The people of Boston became more and more antipathetic to all royal authority, as the "idea whose hour has come" of American independence gathered force; and Governor Bernard, pleasing neither side, was recalled to England in 1769, amid rude rejoicing in Boston. "His position was an impossible one!" said his descendant, opening for our view King George's secret instructions to the governor to give no quarter to these rebels on pain of his personal displeasure.

The Maine estate was confiscated by the new Commonwealth of Massachusetts. Interestingly enough, the eldest son, John Bernard, clearly an individualist, returned to America during the siege of Boston in 1776, landed at Salem, and made his way to Maine. General Rufus Putnam found him on Passamaquoddy Bay in 1784, "in a small hut of his own building, with only a little dog for companion," keeping an eye on his father's estate, which had been bequeathed to him. In 1785, John, now a citizen of Bath, Maine, petitioned Massachusetts for Mount Desert and was awarded half of it for "the propriety of his political conduct." The next year, however, he mortgaged it and returned to England and the baronetcy.[20]

Probably Sir Francis's most important contribution to cartography was to have careful surveys made of the roads from Boston to Saint George's Fort in Maine, on a one-inch to one-mile scale, and from Boston westward to Albany, New York, on a one-inch to two-thirds-mile scale. It was along part of this Albany to Boston road that the American rebels dragged the heavy cannon captured at Fort Ticonderoga that, set up on Dorchester Heights, forced General Howe's evacuation of Boston in 1776. The crossroads, meetinghouses, churches, and taverns with their owners' names are marked. It would have been a notable example of pub-crawling if one had stopped for a drink at each of the taverns on the Boston-Albany route, for they occur every five or six miles; taverns, however, served other important purposes. This was long before the plank roads of the 1830s were devised, and in deep snow or summer rains the going must have been slow and the need for overnight accommodation or replacement of horses frequent. No route maps as detailed as these, except for two short New Jersey road maps, are known for any other section of the eastern seaboard until those of Christopher Colles in 1789.[21] Colles's route maps do not overlap the Bernard maps; he gives the road on the scale of one-inch to one-mile from Albany, New York, to Williamsburg, Virginia, with a few side routes.

The Vermont area appears on some maps as part of New Hampshire and on others as part of New York during the third quarter of the eighteenth century, since both provinces claimed it. The king, in council, decreed in favor of New York on 20 July 1764, and thereafter maps of the Province of New York, like those of C. J. Sauthier, frequently included the territory eastward to the New Hampshire boundary on the Connecticut River. Before then, Samuel Langdon produced a large-scale "Accurate Map of His Majesty's Province of New Hampshire," including the Vermont area and "exhibiting those parts of his majesty's northern colonies which lie exposed to the encroachments of the French in Canada," which shows English and French grants, forts, settlements, roads, with historical notes and relief by hachures.[22] Another manuscript map, unsigned and somewhat smaller, covering the same area and "taken from attested plans . . . and accurate observations," is also dated 1756.[23] In 1768 Thomas Jefferys included in his *General Topography of North America* a map of New Hampshire, dated 21 October 1761, by Colonel Joseph Blanchard and Samuel Langdon.[24] An untitled manuscript map by Gerard Bancker made about 1775 and a fine "Chorographical Map of the Northern Department" by Bernard Romans, published in New Haven about 1777, show independent and new information, with the location of many grants.[25] Romans's map is designed as a companion to his 1776 map of the Southern Department. It is considered to be the first map with the name Vermont on it, and it differentiates the New Hampshire grants by Governor Benning Wentworth after 1749 from those of New York governors after 1764.

The first printed general map of New England to be published in the eighteenth century appeared in Cotton Mather's *Ecclesiastical History of New England,* or, as it is more commonly called, *Magnalia Christi Americana* (London, 1702 (fig. 18) .[26] Probably misled by the reference on the title page of the first book to "An Ecclesiastical Map of the Country," both Dr. Douglass, as early as 1749, and Lloyd Brown, as late as 1940, assumed that the map was by Cotton Mather.[27] But the map Mather referred to on the title page is not cartographical at all, but ecclesiastical; it is a list of ministers in each county of the three British colonies in New England, given in the First Book, chapter 7, under the title, "A Map of the Country." If Mather intended to locate these ministries on a map, he did not do so; nor did he make this map. For the source of the map in the *Magnalia* one needs to look elsewhere in the complicated seventeenth-century cartography of New England.

The first British map of New England is Smith's 1616 chart,[28] followed eighteen years later by William Wood's woodcut, "The South part of New-England . . . 1634."[29] Hubbard's *Narrative,* 1677, has two similar maps, one made in Boston ("the White Hills map") by John Foster, and the other, the same year, in London ("the Wine Hills map"). Both names refer to the present White Mountains. David Woodward of the Newberry Library has recently solved the intricate differences, repeatedly analyzed, between these two plates by showing conclusively the priority of the Boston plate.[30] None of these seems to be the direct prototype of the *Magnalia* map. Before Hubbard's work, and more widely influential, was a series of Dutch maps of the New Netherlands and New England area by De Laet (1630), Jansson (ca. 1651), Visscher (ca. 1665), and Allardt (ca. 1674), which showed increasing knowledge and detail; these were the basis, direct or indirect, of a number of English maps by William Berry, Robert Daniel, John Seller, and other mapmakers in the last quarter of the seventeenth century that depicted the New England area.[31] Justin Winsor suggests that part of the *Magnalia* map was derived from one entitled "A Description of New England," published by John Seller about 1676.[32] Actually the *Magnalia* map, south of the latitude of Cape Ann, follows in detail, for the area given, Thornton-Morden-Lea's "A New Map of New England, New York New Jersey Pensilvania Maryland and Virginia" (ca. 1685; figs. 18–19). Mrs. Newman Hall of Washington, D. C., who is making a careful study of the topography and toponymy of the *Magnalia* map, suggests that its direct source may be another map by Thornton-Morden-Lea of which the only known copy is in the Bibliothèque nationale.[33] This large map apparently incorporates the plate of the 1685 Thornton-Morden-Lea map unchanged, with title and neat

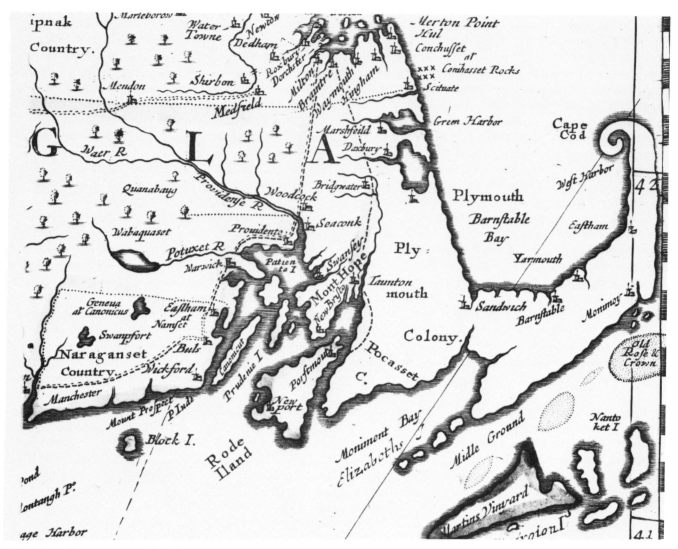

Figure 18 John Thornton et alia [detail] from *A New Map of New England . . .* (London, ca. 1685) [separately published]. Cour- tesy of the Ayer Collection, the Newberry Library.

32

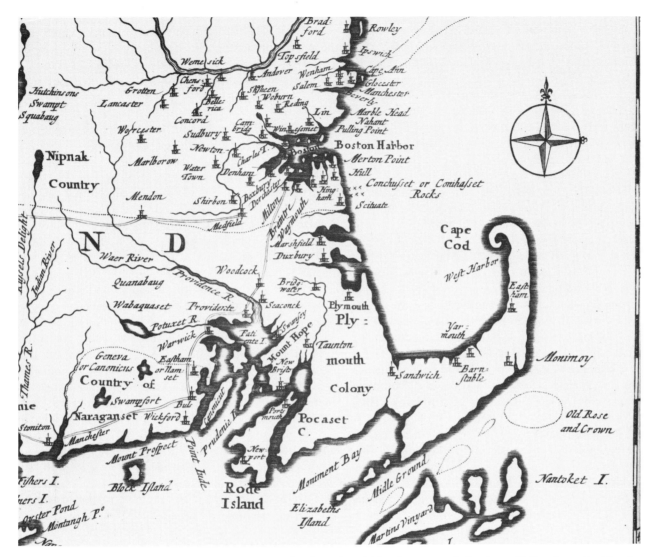

Figure 19 [Detail] from *An Exact Mapp of New England and New York,* in: C. Mather, *Magnalia Christi Americana* (London, 1702). Courtesy of the Ayer Collection, the Newberry Library.

33

lines; but it also has above it the New England area north of Cape Ann to Casco Bay and Falmouth, as does the *Magnalia* map. The Bibliothèque nationale copy is dedicated to James II; this dates it before 1688.

Many British general maps of New England appeared in the first half of the eighteenth century, but with little additional improvement until the work of Dr. William Douglass, a celebrated Boston physician, in the middle of the century. He came to Boston as a young man; in the small-pox epidemic of 1721 he was opposed to inoculation and attacked it in a work published in 1722. The "honest and downright Dr. Douglass" later revised his judgment, however, and gave a more favorable opinion in his later writings.[34]

Of the *Magnalia* map he wrote in his *A Summary, Historical and Political, Of the . . . British Settlements in North America,* "Dr. Cotton Mather's Map of New-England, New-York, Jersies, and Pensylvania, is composed from some old rough Draughts of the first Discoverers, with obsolete Names not known at this Time, and has scarce any Resemblance of the Country; it may be called a very erroneous, antiquated Map."[35] Dr. Douglass died in 1752, and the following year appeared his "A Plan of the British Dominions of New England in North America," engraved by R. W. Seale and published in London in 1753. It is an impressive achievement, both in execution and in its detailed information; it is, however, of such extreme rarity that it has remained almost unknown. Dr. Douglass was not a historian or cartographer. His *Summary* is a collection of valuable notes rather than an organized history of New England, and his map decreases in accuracy and detail for the less inhabited regions.

Dr. Douglass's map did not bring him the reward of fame it deserved, in part because it was the basis for the more famous map of New England published by Jefferys two years later.[36] This is "A Map of the Most Inhabited Part of New England," which appeared in many atlases of Jefferys and his successors in various editions. The author, whose name does not appear, was Braddock Mead, alias John Green, Jefferys's brilliant but eccentric cartographer and draftsman. Of Mead I shall speak more fully in the

next chapter. This map continued to be the chief authority for the region until after the Revolution and ranks in importance with Mouzon's "North and South Carolina," Fry and Jefferson's "Virginia and Maryland," and Scull's "Pennsylvania."

CANADA

Two major British surveys in North America, surpassing in extent and in technical brilliance any preceding achievement in this country, each involving a large and able staff, were the mapping of the province of Quebec under Murray and the charting of the coast by Des Barres and his associates. Before 1759, Canada was French territory, and most English maps were derivatives of the French. The steadily extending and often notable cartographical work of Samuel de Champlain, Louis Jolliet, Louis de Hennepin, J. B. L. Franquelin, and the Jesuit fathers in the seventeenth century, of Baron Lahontan, Guillaume Delisle, and Nicolas de Fer in the early eighteenth century, and the hydrographical achievements of the Marquis de Chabert and Jacques-Nicolas Bellin in the middle of the century are reflected in the atlases and maps of Wells, Moll, Senex, Popple, Mitchell, and others. At the end of the French and Indian War, however, the English found themselves suddenly in possession of a vast region about which they knew little.[37] When Lord Jeffery Amherst conquered Montreal in 1759, he captured the French maps there. (Many of his maps I examined while they were in a private collection, and I shall return to them in the fourth chapter). Amherst, however, felt the need of a thorough survey of the newly won region and of its approaches to the British colonies to the south. In 1760 he ordered Brigadier General James Murray, the governor of the Province of Quebec, to make a survey of the province. For this task Murray was able to command the services of a group of young engineers of whom several were later to become famous for the brilliance of their work: Samuel Holland, John Montresor, Charles Blaskowitz, William Spry, Lewis Fusier, and others. By 15 November 1761, a year later, Murray was able to announce that the field-work was completed. The hardest part of the work on the map appeared to be over; General Murray found that it had just begun. Bitterness developed between the chief engineers for credit and for consequent advancement in rank, which was notably slow among engineer officers. Montresor was penalized by being assigned a particularly difficult and disagreeable piece of surveying because he had erased Holland's name, leaving only his own, when packages containing maps were prepared for shipment.[38] The outcome was not serious: Samuel Holland later became Surveyor General of the Northern District and Montresor was Chief Royal Engineer in North America during the Revolution.

Murray and his staff succeeded in completing their great undertaking by 1762, in spite of bickering and some underhanded politicking. In its complete form, the "Murray map" measures forty-five by thirty-six feet, on a large scale of one inch to two thousand feet, with every house, church, cemetery, and mill along the riverfront shown. Five copies in slightly varying forms are known to exist; two of them are in the British Museum. In 1767 Samuel Holland made a reduced copy, "A Plan of the Settled Part of Canada made in the years 1760 and 1761 by order of General Murray . . ." Several of the engineers who made the map continued their work in surveys for *The Atlantic Neptune*. "The survey holds an important place in historical cartography," writes Nathaniel Shipton. "It was one of the biggest and most difficult ever undertaken by British mapmakers until then, and a milestone on their rise in eighteenth-century cartography."[39]

Jefferys and Faden published a number of maps and charts of different sections of Canada in the years just before and during the Revolution. One of these is "A New Map of the Province of Quebec . . . By Captain Carver and Other Officers in His Majesty's Service," published in William Faden's *North American Atlas* in 1777. Carver's map shows the boundaries of the Province of Quebec according to the Royal Proclamation of 7 October 1763. It does not lay claim to originality; the title includes a statement that it is based on French surveys of the region, "connected with those made after the War." It does, however, show clearly the boundaries of the province, and the details of the main map as well as the in-

sets are finely executed. The inset plan of the Saint Lawrence shows the French feudal estates on the Saint Lawrence and such roads as existed on either side of the river and its tributaries.

Jonathan Carver knew much of the area shown on the maps, since he fought throughout the French and Indian War and had been promoted to captain of a provincial company in 1760 after serving under Wolfe at Quebec. He was living in Boston before leaving for England in 1769. His name is associated with three other maps, two of which appeared in his well-known *Travels Through the Interior Parts of North America*, published in London in 1778: "A New Map of North America from the Latest Discoveries" and "A Plan of Captain Carver's Travels in the Interior Parts of North America in 1766 and 1767." Carver has been called "the first traveller of English speech to explore any part of the interior west of the Mississippi River."[40] His maps purport to show his discoveries; they do not add much to the knowledge of the interior, although his general map is a good representation of what was known or thought to be known, and shows the legendary River of the West that flows to Drake's New Albion and the Strait of Anian. Carver in his *Travels* calls this river the Oregon, the first time that name is used. He says he learned much from maps by Dakotah Indians drawn on birchbark in charcoal; but much of his information came from French and other sources. His career, both as a writer and a married man, was not without darker shades. He did not return to Boston, where he had left a family unprovided for; he married again and raised another family in London. He apparently hoped to be appointed Superintendent of Indian Affairs in America, but in this, if not in his marital achievements, he was disappointed.

THE TRANS-ALLEGHENY REGION

The Trans-Allegheny region, including the Mississippi River and its tributaries, was dominated by the French during the first half of the eighteenth century. French explorers descended and ascended its rivers, French priests established its missions, French traders bought its furs from the Indians, and French geographers made its maps. Meanwhile the English were curious and began to extend the tentacles of their knowledge beyond the mountain ranges. It was not only the South Carolina expansionists like Nairne whose ambitions stretched toward the Mississippi. Hunters and trappers began to explore the rivers flowing westward from the Appalachian ranges; frontiersmen and land speculators looked for fertile valleys for settlements. Men of wider vision like Lewis Evans and Indian scouts like Christopher Gist argued that the control of the Ohio Valley was a matter of life and death for the British colonies. As the tensions of impending war increased and the strategic importance of learning the best military routes became clear, the information needed was gathered and put down on maps.

At first, only French maps were available. In 1724 Cadwallader Colden, the first Surveyor General of the Province of New York, published "A Map of the Countery of the Five Nations" that was an exact copy of a small section of Delisle's map "Louisiane et du Cours du Mississipi" (1718), with the addition of a few portages and trails. Colden's map extends from the Hudson to the eastern shore of Lake Michigan.[41] John Patten, a captive of the French in 1750–51, made a map of the Ohio River that has been studied intensively and ably in Howard Eavenson's *Map Maker and Indian Traders*.[42] About 1753 George Mercer, agent for the Ohio Land Company, made a map to identify the boundaries of grant lands and to show a suitable place for the company's headquarters; the legends explain his choice for a fort on a rocky cliff.[43] This map, now in the Public Record Office, was responsible for the building site of Fort Prince George the next year, 1754. Another map in the Public Record Office is by George Washington, with notes on it in his own hand, showing the pioneering route that he took with Christopher Gist in 1754 on his way to Fort Le Boeuf and a conference with the French there.[44] It accompanied Governor Dinwiddie's letter to Lord Haldeman. The map in Washington's *Journal*, published in 1754 in London by Thomas Jefferys, a "map of the Western parts . . . of Virginia," uses Washington's map but covers a much wider area.[45]

"Captain Snow's Scetch of the Country by himself and the best accounts he could receive from the Indian Traders 1754" has fewer details and is less accurate geographically than the best maps of the period. But it has numerous notes showing the location of forts, paths, Indian attacks, and encounters with the French. The original manuscript is in the Library of Congress.

If the Northern Indian District west of Pennsylvania and New York has no counterparts to the great maps of the southeast made by John Stuart at the end of the Colonial period, it is due to the attitude of the Superintendent of Indian Affairs for the Western Department, Sir William Johnson. There is no evidence that he ever planned to produce any maps on the order of those made by Purcell. Nor was there any surveyor general for the north to equal De Brahm or his deputy surveyor, Bernard Romans, in the south. Sir William was able in his dealings with the Indian tribes; but the maps made after the Treaty of Fort Stanwix are notably lacking in detail and cartographic skill.

The best British map of the region south of the Great Lakes was made by Captain Thomas Hutchins and published in 1778. An earlier dual map of the Ohio Valley, by Captain Hutchins, published in 1766, is a beautiful piece of draftsmanship. His "New Map of the Western Parts of Virginia . . . 1778" accompanied his "A Topographical Description" and was the culminating effort of his years as a captain of the 60th Regiment of Foot.[46] The map has many interesting legends concerning minerals, grains, wild life, and geography; especially appropriate here is his note on Lake Michigan, "a vast Collection of Fresh Water . . . In Lake Michigan is great Plenty of Fish, particularly Trout . . . some have been taken . . . which have Weighed upwards of 90 pounds."

I have endeavored to trace in these first two essays the development of British colonial cartography from the local to the provincial to the regional. For the first half of the century, the best general maps of the British dominions are those of Moll in *The World Described* (1709–20), Popple (1733), and Mitchell (1755). At the outbreak of the Revolution, two excellent printed maps, already mentioned, attempted to incorporate the results of surveys in the third quarter of the century: Bernard Romans's for the southern provinces and Thomas Pownall's from Virginia to Canada. There had been a tremendous advance during the century in the techniques of surveying and in the accuracy and detail of the maps; but at least in the North this appears more clearly in the local than in the general maps. In the early years the maps were still primarily descriptive, often based on the observations of travelers, traders, and Indian agents as they explored new territory. By the middle of the century trained surveyors such as De Brahm and Romans in the South and Evans, Nicholas and William Scull, Sauthier, Holland, and Montresor in the North were at work with able assistants they had taught. With improved instruments at their disposal they had begun triangulation surveys, especially along the coast; they had greatly increased the number of astronomical observations throughout the colonies for latitude and, occasionally, for longitude. The great outburst of cartographic activity during the Revolution will form part of the account in the following two essays.

Charting the Coast
Simple Beginnings
to Professional Accomplishment

Coastal charting is very different from mapping the topography of the interior of a country. It requires different skills such as knowledge of navigation and of stream and tidal movements and different equipment such as small surveying boats and sounding leads. If it has some of the same motivations, such as exploration, establishing trade, and finding suitable places for settlement, it also has other objectives and other emphases.

Marine charts may not be as old as land and property maps such as those with cuneiform inscriptions found in Asia Minor; but the portolan charts of the Mediterranean made in the Middle Ages are among the oldest surviving attempts to delineate the earth's surface. If you are crossing the Atlantic to explore the New World, you make charts of the coast first before you map the interior: the maps of Cosa, Ribero, Juan Vespucci, and Verrazzano gave the outlines of the North American shoreline long before there were inland maps.

Coastal charting is tremendously concerned with life-and-death safety, especially along hazardous and complicated coasts like Maine, and the Carolina Outer Banks ("The Graveyard of the Atlantic"); the "showldes" off Cape Cod are potentially as dangerous as the rocks off Maine. One admires the enormous courage of those who first approached the shores or who went in to chart them. Rosier, on Waymouth's 1604 exploration of the New England coast, gives with great vividness the dismay of the voyagers when suddenly finding themselves in shoal water with the "whitish sandy cliffe" of the eastern end of Nantucket rising up ahead. He adds that they found their "sea charts very false, putting land where none is."[1] The dangers of unknown coasts were fewer at the beginning of the eighteenth century than one hundred years earlier; but there were many vaguely charted stretches along the shore and many more incorrectly charted. "Houracanes" and storms, winds and currents, were special adversaries to the marine surveyor as well as to those who profited by his work.

In the eighteenth century, rivers were the roads to settlements, to new regions, in trade and in war. In the beginning there were no roads and a waterway was thought essential: it was necessary for the Virginia planter to get his tobacco from his Pamunkey landing to Hampton Roads, the Carolina indigo and rice grower his produce from his Edisto Island plantation to Charles Town, and the English merchant his furs from Albany down the Hudson River to New York. Therefore, river charts were correspondingly important.

EARLY ENGLISH CHARTMAKERS: DUDLEY AND SELLER

The relationship of Britain to her colonies across the sea in North America demanded adequate coastal charting to reach harbor safely, to facilitate the trade that was the chief function of the colonies, and to prosecute the wars that arose. During most of the seventeenth century, charts of the North American coast were Dutch. Champlain had made a few detailed charts of harbors in New England and Nova Scotia, with some soundings; whether these were known to or used by English mariners is questionable. The first soundings on any map of the southern coast were by Robert Dudley in his *Dell' Arcano del Mare,* made in Italy in 1646–47.[2] His source is unknown; in any case, the soundings are as inaccurate as his coastline. Dudley made his map of the coast south of Cape Fear by superimposing the coastal nomenclature of Hondius's map of 1606 on an earlier Spanish map, so that on this coast, rich in names of rivers, capes, and bays, the number of place-names is markedly increased. For many features, he gives both French and Spanish names and draws in extra rivers to accommodate both. Dudley's *Dell' Arcano del Mare* was the first atlas based on Mercator's projection; it is an important and beautiful work finely engraved by Lucini. Some of his charts, however, thoroughly out of date, must have been bewildering to the navigator who actually tried to use them. His chart based on White's and Smith's "Virginia" was better than his more general chart of the Southeast coast.

JOHN SELLER

As the number of colonies increased during the last half of the seventeenth century, the urgent needs of English merchantile trade and of the Royal Navy stimulated the publication of English maps and charts. For a hundred years the Dutch had almost monopolized the field; British seamen depended on the Netherlands for charts even of their own coasts. When an enterprising maker of compasses, nautical instruments, and maps, John Seller, produced an English sea atlas, he was given the title "Hydrographer in Ordinary" to the king. Under letters patent in 1671 he was protected from pirate printers of his maps or importation of copies from across the channel, "either under the name of Dutch Waggoners or any other name whatsoever," within the term of thirty years. This protection was, however, ineffective; Samuel Pepys, the Secretary of the Admiralty, wrote that the Dutch were copying and selling Seller's maps in England. The practice worked both ways, however, for Seller had found that it was easier to promise an enlarged sea atlas than to produce one, and Pepys notes elsewhere that Seller had gone to Holland and "bought the old worn Dutch plates for old copper and had them refreshed in several places and used them in his pretended books."[3]

If, however, Seller bought copper plates abroad and used them or the information on them, he also added new information to them. He derived his fine chart of New Jersey from a Dutch map; but he based his delineation of the Delaware River and Bay on Herrman's "Virginia," which he himself had published in 1673. On his "A Chart of the West Indies" (1675), he copied the standardized Dutch West Indian paskaerts or navigational charts; but he also incorporated many details from John Ogilby's "First Lords Proprietors' Map" of 1672 for the Carolina region.

In 1671, Seller published his *English Pilot: The First Book,* which had marine charts of the northern coasts of Europe. Shortly afterward he began to publish other sea atlases, with various titles, which had some charts of the American coast. Seller soon turned over his project of a multivolumed *English Pilot* to a nautical publisher, William Fisher. Fisher, in turn, joined forces with John Thornton, a cartographer who had drawn charts for Seller. Together they published *The Fourth Book* of the *English Pilot* in 1689, the first significant collection of charts exclusively of the American coasts to be published in England.[4] For British trading in North America and for the colonists there, the publication of *The English Pilot: The Fourth Book* must have been a godsend. For the first time an English sea atlas presented charts of the whole eastern seacoast of North America. To modern eyes the charts are crude and sparse of detail; but to the navigator of American waters in that period, it was his Bible. Whatever its shortcomings, there was really no substitute, no real competitor, for over sixty years.

THE THAMES SCHOOL

John Thornton, copublisher of *The Fourth Book* and prolific chartmaker himself, had learned his trade as an apprentice (1656–64) to John Burston, a member of the Drapers' Company. Burston in turn had been an apprentice of Nicholas Comberford, producer of many charts, also of the Drapers' Company, who trained at least five other apprentices in his shop. Professor T. R. Smith of the University of Kansas has shown that Comberford was an important figure in a long series of Drapers' Company platmakers who worked in London during the seventeenth century and after.[5] The existence of the Draper Company chartmakers and their identity have been in the process of discovery for the past fifteen years, with a number of persons on both sides of the Atlantic contributing to the study. Miss Jeannette Black of the John Carter Brown Library has suggested the name, now generally accepted, of the "Thames School" for these chartmakers, since the shops of a number were near the river's edge, near the Tower, in Radcliffe on the north, or in Rederiffe on the south bank. That the Drapers' Company, clothmakers, should have taken chartmakers under their wing may seem strange. The Drapers and other livery companies, however, had lost their original character of guilds by the sixteenth century and had developed some characteristics of fraternal and benevolent organizations.

Much more investigation remains to be done concerning the membership and occupations in the companies at this time; it may be purely accidental that so many platmakers were Drapers. John Seller, who constructed manuscript charts similar in style to those of the Thames School, was a member of the Merchant Tailors, although he was neither a tailor nor a haberdasher. While it is the master-apprentice relationship that truly characterizes the members of the Thames School, there is a marked similarity in style and decoration, maintained over seventy years, that suggests a cartographic genre.

Comberford in 1657 made a map of the Pamlico-Albemarle Sound area that has resulted in the identification of what may have been the first permanent settlement in North Carolina. The map is now in the New York Public Library. On Comberford's map is a house bearing Captain Nathaniel Batts's name (fig. 20). My finding of this detail had an amusing result, for the people of Bertie County, North Carolina, became much excited when I reported this in the *American Historical Review*[6] and in 1957 the county put on a tercentenary celebration of the first permanent English settlement south of Virginia, to which I was invited. It was sponsored by the Chamber of Commerce, and I went with some apprehension that I might be put on a float surrounded by bathing beauties; such are the strange exigencies in which the historian of

cartography occasionally finds himself. Near the site of Batts's house the State of North Carolina has erected a historical marker to him. In 1672 George Fox, the founder of the Society of Friends, visited Batts, whom he called "a rude and desperate man, . . . who had been Governor of Roan-oak."[7]

To return to the Thames School, John Thornton's apprentice, Joel Gascoyne, made "The Second Lords Proprietors' Map of Carolina" in 1682 and also drew the great Maurice Mathews wall map of Carolina now in the British Museum.[8] Gascoyne had an apprentice named John Friend. In October 1969, Dr. Wallis and I found in the vaults of Chatsworth House fifteen vellum charts by John Friend, dated from 1703 to 1708. They are all in the characteristic style of the Thames School. The Chatsworth portolan charts form a series that depicts the entire coast of Asia from the Persian Gulf to Korea. They overlap the maps in *The English Pilot: The Third Book* but are not duplicates.

In the Public Record Office are two complementary maps of the southern and northern parts of Chesapeake Bay.[9] They were made about 1681 and, while they have no identifying signature, they have so many characteristics of the Thames School of portolan chartmakers that they were apparently made by one trained in the school.

SOUTHACK, HOXTON, AND WIMBLE

The deficiencies of *The English Pilot* were serious and became more obvious as local mariners and pilots accumulated knowledge, discovered hazards, entered new information in their logbooks, and made new charts. The publishers of *The Fourth Book* were slow to improve their maps; they had a monopoly and, impervious to the loss of ships and life, they did not wish to undergo the expense of new plates. Their plates occasionally had corrections made on them; some charts were omitted from later editions; and sometimes in their place new ones were added, which they made or bought from other publishers.[10] But changes were made with reluctance; there are charts in the last edition that had first appeared a hundred and four years before. The chart of the Carolina

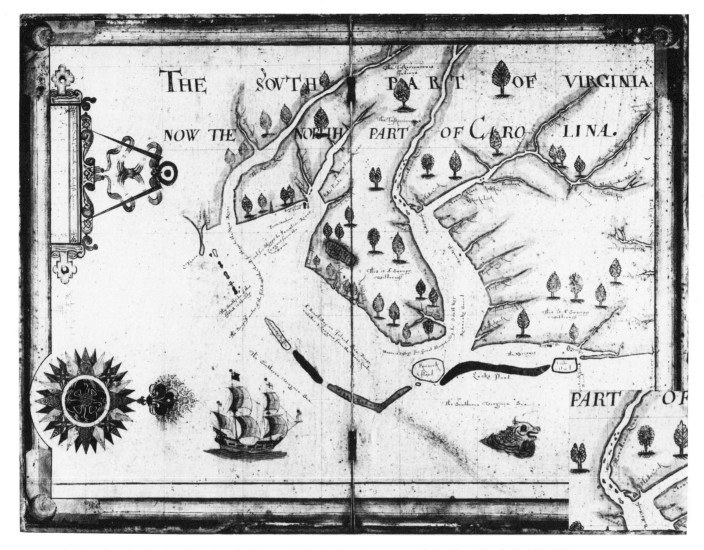

Figure 20 Nicholas Comberford, *The Sovth Part of Virginia*   Courtesy of the New York Public Library.
Now the North Part of Carolina (1657) [and detail] [manuscript].

coast well illustrates their practice, for it depicts an area especially dangerous both to local navigation and the heavy coastal trade from the North to the West Indies. The chart appeared in the first edition of 1689 and was used for the next sixty years, with no change or corrections on the plate. Yet two of the inlets along the Outer Banks were closed at least by the early years of the eighteenth century; Roanoke Inlet was wrongly placed; and Cape Lookout, which jutted out into the Atlantic in mid-sixteenth-century Spanish charts with the ominous name of "terra falgar," was shown as scarcely a protuberance on the coast. It is a striking example of the low standards of mapmaking in England in the first half of the eighteenth century. This can be illustrated further by an examination of the relationship to Mount and Page of three men, Cyprian Southack, William Hoxton, and James Wimble, who made charts of the American coast for the publishers of *The Fourth Book* during most of the eighteenth century, but whose valuable work was not included in it.

Cyprian Southack is perhaps best known for his *New England Coasting Pilot,* an atlas in eight sheets, published sometime between 1720 and 1734. He came to Massachusetts in 1685 as a young man of twenty-three, an officer in the coast guard that patrolled the New England shores. He searched for pirates, rescued crews wrecked or attacked by Indians or French, and surveyed and made charts of the coast from the Gulf of Saint Lawrence to New York. Twenty or more maps, printed or in manuscript, survive to show Captain Southack's industry in charting the coast of New England, Nova Scotia, and the Gulf of Saint Lawrence, and in making his cartographic labors available for others.[11] His first chart was "A Dravght of Boston Harbour . . . 1694."[12] He proudly recalls in a notice attached to his *Coasting Pilot* that in 1694 for his many services he was granted permission to kiss King William's hand and received fifty pounds to buy a gold chain and medal. About 1710 he constructed a large and handsome chart of the Gulf and the River of Saint Lawrence, now in the Public Record Office.[13] He published in Boston "A New Chart of the British Empire in America," engraved by Francis Dewing in 1717, the first

map engraved on copper in North America.[14] Its geographical information is very uneven; it shows a strong political anti-French motivation. For the coast north of Boston, Southack's charts were superior to those in *The English Pilot: The Fourth Book,* especially for the soundings along the coast and in the harbors and for the navigational information in the numerous legends on *The Coasting Pilot.* The Massachusetts Historical Society Library has the rare first edition of the 1689 *Fourth Book.* It is Southack's own copy with rough drafts of letters written on the back of the maps. In the Naval Library in London I found an unrecorded manuscript by Southack, "The Harbour & Islands of Canso part of the boundaries of Nova Scotia," signed Cyprian Southack, 1718.[15] It is similar to a chart engraved by Francis Dewing of Boston in 1720, "The Harbour and islands of Canso." The same year, 1720, Emanuel Bowen, a London engraver and publisher, made a copy of Southack's "The Harbor of Casco and islands adjacent" that appeared in *The English Pilot: The Fourth Book* in 1721 and in subsequent editions for seventy-three years, without improvement. In the 1775 and later editions, J. Mount, T. Page, and W. Mount added to *The Fourth Book* "A Map of the Coast of New England"; it was a simplified, one-plate chart of a section of Southack's *Coasting Pilot* acquired from John Senex, who had made it about 1744. *The Coasting Pilot* was a great advance over previous charts of the area.[16] But it was not until 1775, at a time when Cook, Holland, and Des Barres were producing the results of their notable surveys, that the publishers of *The English Pilot* at last began to include a section of Southack's *Coasting Pilot,* but without his useful annotations that they had obtained thirty years before.

William Hoxton's "This Mapp of the Bay of Chesepeack, with the Rivers Potomack, Potapsco North East, and part of Chester," was engraved and published in London by B. Betts in 1735; W. Mount and T. Page reissued it in 1750. "It is impossible to write on the subject of American contributions to navigation without mentioning this large and detailed chart," said Lawrence Wroth, who pointed out that in a long legend Hoxton made a table on the course, strength, and limits of the Gulf Stream. Hoxton anticipated the work of Benjamin Franklin and De Brahm on its significance in transatlantic navigation (to and from Virginia).[17] Hoxton's map, published with Mount and Page's imprint in 1750, was clearly superior in accuracy and navigation aids to the old Mount and Page map of the same area that continued to appear in *The English Pilot.*

For reasons of expense, commercial chartmakers depended on volunteered information and charts from captains whose trade or official duties provided them with constant opportunity for soundings, observation of shoals and harbor passages, and correction of existing charts. These men were eager to share their knowledge, and their comments were sometimes included in the sailing directions that accompanied the charts; for example, following a chart of Cape Fear, Joseph Speer, in his "West India Pilot," notes: "These directions may be depended on, and are inserted by the desire of Capt. Thomas Potts, in this trade, for the instruction of those who may come on this coast." *The English Pilot* would have been far more helpful to navigators if it had been updated more frequently and conscientiously and corrected from information on new charts with revised instructions. In the latter years of the century greatly increased competition caused the publishers to make changes and include additional maps.

A revealing account of how these early coastal charts originated and came into print may be found in the activities and background of a Boston innkeeper, distiller, trader, privateer, and sea captain, Captain James Wimble. His map of the North Carolina coast, published by Mount and Page in 1738, is the best hydrographic chart of the region before the closing years of the century. Twelve years ago a chance comment by an Oxford student, now a well-known novelist, directed me to a manuscript map of the Carolina coast in the Rawlinson Collection of the Bodleian Library.[18] It was made by Wimble in 1733; though crude in execution, it had more detail than any other chart of the area up to that period, and even gave some data omitted on his printed map five years later. I undertook to learn what I could about this unknown captain and his motivations for this pioneering endeavor in coastal charting. After Mrs.

Cumming and I journeyed to small, dusty courthouses along the Carolina coast, to the Massachusetts Historical Society, and to the Institute for Historical Research (University of London) and the Public Record Office, his story emerged in surprising detail from letters written by and about him, deeds, land grants, charts, and the august files of the Lords of Trade and Plantations.[19]

Wimble was born in Hastings, Sussex, in 1696, and trained as a seaman. His lack of early academic education reveals itself in the most fearsome spelling ever encountered in many years of early map transcription. (One is tempted to wonder whether the failure of his patron, the Duke of Newcastle, to do more for him than he did was occasioned by his being frequently addressed by Wimble as "His Grace the DUCK of Newcastle.") The family of ten fell on hard times, and James, having built a small boat with the aid of friends, sailed in 1718 to the West Indies to seek his fortune.

Wimble's activities in the next few years were varied. Based at first in the Bahamas, he began to trade with the mainland colonies, especially the Carolinas. Soon he was buying land along the North Carolina coast. He was also drawn to Boston, where he set up a distillery, a practical outlet for one engaged in the rum and sugar trade with the West Indies. He bought land in Boston and married the daughter of a substantial citizen. From there he continued to trade with the Carolinas and with the West Indies, although Spanish hostility was making this ever more hazardous, and in 1728 he was "taken" by a Spanish privateer, losing both sloop and cargo.

In 1730–31, Wimble bought a large tract of land in North Carolina on the Cape Fear River, where he later became one of the three cofounders of the town of Wilmington. He had a brigantine, the *Rebecca,* built for him there. In all his voyages along the Carolina coast in pursuit of these interests, he was severely, dangerously handicapped by the inaccuracies of *The Fourth Book of The English Pilot,* the only charts available to him. He began to make his own, which he completed in 1733.

In 1731, however, Wimble had suffered a serious disaster. He had loaded the *Rebecca* with "Turpentine, Tarr and Lumber," and sailed from Cape Fear for Boston. Meeting a violent storm, which greatly "damnified the vessel," as he says, he was finally forced to put in to New Province to stop the leaks. The governor of the Bahamas at this time was Woodes Rogers, a forceful and arbitrary man with much to contend with from pirates and Spaniards. He refused to let Wimble sail, and impressed him and his crew and vessel to carry timber and to guard the salt ponds against marauders. Wimble's vigorous protests availing nothing, he was engaged in this mission when another terrible hurricane struck, and he barely escaped with his life, losing his ship, his investment, and much of his substance. He immediately began a long, fruitless suit to the Lords of Trade and Plantations to recover damages, the files of which, with many of Wimble's own statements and letters, are in the Public Record Office. He collected witnesses, he traveled to London, he invoked his patron, only to be told in the end that he had not proved his case. He never was able to believe, as so many Americans were beginning to do, that the government of England would use him but not serve him; the last we know of him is that he lost an arm fighting for his king in the War of Jenkins' Ear, still planning a new appeal for his deserts in the loss of his brigantine.

All the while that Wimble was pressing his suit, however, he was briskly engaged in other activities: coastal trading, selling lots on Cape Fear as the new town of Wilmington took shape, fathering his family in Boston, and making his coastal charts. He may have hoped to realize something from the sale of these; he undoubtedly hoped for increased favor with the Duke of Newcastle, to whom he dedicated the 1738 map; but certainly his strongest motivation was the great need for safety, his own and others'. Dangerous as navigation was along the North Carolina Outer Banks and treacherous as shoals and sandbanks were to inland navigation, no chart of the region had yet been printed that approached in adequacy the charts in existence for Boston, New York, the Chesapeake, or the South Carolina coast. So Wimble, in spite of his difficulties, applied

his extensive knowledge of the best anchorages and of the depths of water in the sounds and up the rivers all the way from Albemarle Sound to Cape Fear. This he had gained by trading at great risk up and down the sounds and rivers with plantation owners who loaded their produce aboard ships from their own or nearby landings.

Wimble evidently hoped that his manuscript map, sent to England in 1733, would be substituted for the inadequate chart of Carolina in *The English Pilot*. He wrote on the back of the map, to Captain Cullen, asking that he dedicate it to, and have it printed by, William Mount, senior partner in the firm that published *The English Pilot*. It was never printed, possibly because of the appearance of the Mosely map of North Carolina in that year; how the manuscript reached the Rawlinson Collection in the Bodleian is not known. It is strictly hydrographic, giving soundings, anchorages, and passageways for ships, piraguas (two-masted, undecked, flat-bottomed boats), and canoes. It reflected his own land purchases on Albemarle Sound and on the Cape Fear River, where in "New Carthage," later to become Wilmington, he draws an appealing picture of his own house, labeled "Wimbleton Castele" and flying a large British flag. We have evidence in 1739 that Wimble had built a house on a high bluff overlooking a river. That it was a "castele" is improbable. On C. J. Sauthier's manuscript map of Wilmington, 1769, a square house, thirty-six feet by thirty-six feet, with outbuildings and gardens, stands on Wimble's lot, and was probably occupied by his son, who advertised for sale, in 1766, "a very genteel new Chaise." The penniless lad who built his own boat in Sussex had come a long way.

Wimble's printed map of 1738 remained the best hydrographic chart of the North Carolina coast till nearly the end of the century. He must have continued his surveys after 1733, since this later map gives more detailed and sometimes different information along the coasts and up the estuaries and rivers. It adds the names and locations of various settlers along the waterways. While Wimble's own name appears on the map seven times (four as a property owner),

Wimble Shoals, off Cape Hatteras, is the only eponymous feature remaining today. This 1738 chart, dedicated to "His Grace Thomas Hollis Pelham Duke of Newcastle," was printed by Mount and sold in both London and Boston. But for twenty more years Mount and Page continued absurdly—one might say irresponsibly—to print the erroneous seventeenth-century chart of Carolina in *The English Pilot: The Fourth Book.*

JEFFERYS AND GREEN (MEAD)

In the middle of the eighteenth century, new forces caused the long preeminence of the publishers of *The English Pilot* to wane. One was the printing of charts by seamen like Southack and Wimble that were better than those in *The English Pilot.* Another was the emergence of new printers willing and able to produce better charts. A third was the rapid improvement of scientific methods of marine surveying. It was, however, the threat of oncoming war, with the resultant insistence on greater accuracy in coastal charting, that gave an accelerated impetus to the change.

Chief among the new publishers was Thomas Jefferys, who became geographer to the Prince of Wales in 1746 and in 1760 to the king. He was the leading British chart and mapmaker of his day, and his work contributed toward making London the "universal centre of cartographic progress."[20] An engraver as well as a publisher, he turned out an impressive number of maps and charts, many of them of the North American continent. He published so many of these that he gathered them into atlases: *A General Topography of North America* (London 1768), *American Atlas* (London 1775), and *The West India Atlas* (London 1775), the last two appearing posthumously. With William Faden, his successor, he produced the most considerable body of North American maps published commercially in the century.

The genius behind Jefferys in his shop was a brilliant man who at this time went by the alias of John Green. He made a great six-sheet map of North and South America (1753), concerning which he said, "The English charts of America being for the general very

inaccurate, I came to a Resolution to publish some New ones for the Use of British navigators." Green had a number of marked characteristics as a cartographer. One was his ability to collect, to analyze the value of, and to use a wide variety of sources; these he acknowledged scrupulously on the maps he designed and even more fully in accompanying remarks. Another outstanding characteristic was his intelligent compilation and careful evaluation of reports on latitudes and longitudes used in the construction of his maps, which he also entered in tables on the face of the maps. These characteristics distinguish and identify such productions as the maps of "Nova Scotia" (1755) and "A Map of the Most Inhabited Parts of New England" (1755). G. R. Crone, former librarian of the Royal Geographical Society, in a cogent analysis of internal evidence, ascribed to John Green several anonymous works that have since been shown conclusively to be his, including "The Construction of Maps and Globes . . . Intermix'd with some necessary Cautions, Helps, and Directions for future Map-Makers, Geographers, and Travellers," published in 1717.[21]

In common with his employer, Green was a vigorous Francophobe; both were ardent supporters of British territorial claims against France in North America. Green was a controversialist. Although acknowledging the preeminence of French geographers and mapmakers, he attacked Bellin with ill-concealed animosity for the faulty determination of latitudes and longitudes on his maps. J. N. Delisle he accused of professional subterfuge and then of outright lying.

It is now known why this man adopted an alias, stayed in the background, and failed to sign a number of his works.[22] The identity of John Green, whose real name was Braddock Mead, is revealed in a letter written by Thomas Jefferys ten years after Mead's suicide, in response to a request from Lord Morton, at that time president of the Royal Society. The extraordinary details of his life appear in two documents, one a manuscript letter in the Ayer Collection of the Newberry Library, the other in a printed pamphlet in the Law Library of the Library of Congress. The manuscript letter is pasted in a copy of Green's *Remarks in Support of the New Chart of North and South America* (London 1753). Jefferys wrote on 17 January 1767 to the Earl of Morton:

My Lord

Agreeable to your Lordships request I sent you a short account of Mr. Green the person whom I employed in many of my Geographical performances & in particular ye chart of North & South America—his true name was Bradock Mead—of Ireland (& his Brother was mayor of Dublin in or about the year 1758 and I believe is still Living) he had university Education & if I mistake not was at Trinity College Dublin at the same time I think with my Friend the Revd Dr. Dobbs who if Living resides in the North of Ireland—I have been in company with them both, since his Death I have seen a flying Printed Sheet—being an Account of his being concerned with some more in running away with an Heiress, he was fortunate enough to gain this Land of Liberty, but some one or more of his companions forfeited their Lives for the Offence. If I am not mistaken this was about the year 1728—when he came into London he took the name of Rogers & served the learned Mr Edward Chambers as an Emanuensis in some part of his elaborate Dictionary. My acquaintance began with him in 1735—when he went by the name of Green & was employed by my late worthy Friend Mr Edw Cave in Translating Du Halde, China, but in the year 38 or thereabouts he differed with Mr Cave & entered into the service of his Antagonist Mr Astley the Publisher of the London Mag. who published his collection of Voyages and Travells—Differing with him also—the work stopt at the end of the fourth Volume having done no more than Asia & Africa—after that he was employed by ye proprietors of the Universal History & last of all by your humble Servant—he was a man of warm passions fond of ye Women & Intrigue, having had a correspondence with —Wife for some years before his Death, he afterwards married her, & threw himself out of window 3 story-high in less than 3 months 1757—this rude sketch is drawn from my memory having no notes, if it should answer the end that your Lordship wanted it will give pleasure to your humble servt

T. Jefferys[23]

The unedifying episode in Braddock Mead's life referred to by Jefferys is given in detail in the thirty-page "Whole Case and Proceedings in Relation to Bridget Reading, an Heiress . . . and of her Pretended Marriage to Braddock Mead," published in London in 1730.[24] This pamphlet reads like an eighteenth-century novel. Mead's actions are reminiscent of those in Smollett's *Roderick Random* and in Richardson's *Clarissa*. A Daniel Reading of London asked his supposed friend, the unscrupulous Daniel Kimberly, to bring back from Dublin his daughter, Bridget Reading, who had been there with a nurse since the death of her mother, and who had come into a small fortune.

"On his arrival in Dublin," says the printed account, Kimberly

getting the child into his Possession, . . . Lodged her with one Mrs. Peters, a Woman of very bad Repute. On the 11th of April, 1728, the Child being then 12 Years and 6 Months, the said Kimberly, one Ambrose, a degraded clergyman, . . . and the said Mrs. Peters, together with Braddock Mead, upwards of 40 years of age, got the Child into a Room amongst them and holding her by Force . . . the pretended Clergyman, pretended to read the Ceremony of Marriage between the Child and Braddock Mead.

Thereupon Kimberly and Mead took Bridget to London, with the intention of securing the inheritance, which by English law belonged to the husband. Reading, however, had them both arrested. Upon Mead's "solemnly protesting he had not lain with the child, and that he was sorry for his Offense," Reading allowed the release of Mead and Kimberly on a £2,000 bond. They, however, brought suit, claiming validity of the marriage and right to Bridget's estate, and attempted to seize the girl in the street. In Ireland a clandestine marrige was a felony punishable by death. Mead also, it appeared, had deserted two former wives. The father sued successfully for extradition of the two to Ireland; Kimberly was arrested, extradited, and hanged. Mead went into hiding, changing his name, and escaped. If Kimberly was the Mephistopheles in the scheme, Mead was certainly not less than a cooperative Faust. His aliases were, however, successful; he continued to live in London as a hack writer; his undoubted learning and abilities were used by others. Mead's contributions to cartography stand out in contrast to the shoddiness of his private life. At at time when the quality and the ethics of map production were at a low ebb in England, he vigorously urged and practiced high standards; in the making of maps and navigational charts he was in advance of his time. For this he deserves due credit.

GASCOIGNE AND COOK

Two less well-known makers of marine charts were Captain John Gascoigne and Lieutenant James Cook. Gascoigne and his brother, Lieutenant James Gascoigne, were assigned to HMS *Alborough* in August 1728 to survey Port Royal in South Carolina. W. E. May in his account of the survey, drawn from original documents, has solved some problems about which I wrote a dozen years ago.[25] Captain Gascoigne had been trained in surveying methods since 1721. In 1722 he observed an eclipse of the moon at Port Royal in Jamaica that enabled him to give with accuracy the longitude of the place; later he pointed out "grave errors and discrepancies evidently competent for his time, the instructions for surveying that he received and the logbook that he kept show how primitive between latitude and longitude in published tables." Although were the methods then employed as compared to those thirty years later. Three boats were sent out from the ship; when a shoal was reached, the boats stopped, anchored, set a buoy, and made observations of the variation of the compass and tidal streams. Those who went ashore used a theodolite and chain. The chart Captain Gascoigne made after six months of work is in the Public Record Office. Years later, Francis Swaine, a marine painter active in the third quarter of the century, made another chart of Port Royal, using Gascoigne's survey; the detail and draftsmanship are of a different order, as one can see by the engraved chart Thomas Jefferys made of it or a copy of it in 1776, nearly fifty years later. Jefferys published another Gascoigne map about the same time, 1776, of "D'Awfoskie River and Harbour in South Carolina." Des Barres, apparently also using the Gascoigne-Swain manuscript but

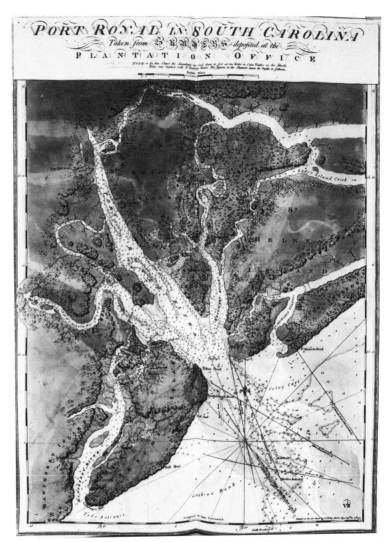

Figure 21 J. F. W. Des Barres [detail] from *Port Royal in South Carolina* (1777), in: *The Atlantic Neptune* (London, 1774–1800?). Courtesy of the Ayer Collection, the Newberry Library.

47

giving new soundings, published his "Port Royal in South Carolina" in 1777. The reproductions of these and other successive maps of the same region show graphically the improvement in cartographic method and detail (figs. 21–24).

Another name that has raised problems is that of James Cook; the confusion here is happily expressed in the title of a forthcoming article by Jeannette Black and R. A. Skelton: "Too Many Cooks." There were three officers in the Royal Navy with the name of James Cook who were in American waters in the 1760s, and all three published charts. Who published which, and which was who? I believe the article quotes me, and I hope Miss Black, who has found out much more about James Cook of South Carolina than I, will allow me to steal a little of her thunder. The famous Captain Cook never was in South Carolina, and another James is known for a chart of Fowey Harbor; our James Cook, who made charts of Halifax, of Port Royal (see fig. 24), and parts of the Gulf Coast, also made the most accurate maps of the Province of South Carolina before the Revolution. His printed map of Carolina appeared in 1773. He had lost his commission in the Navy in 1765, apparently from overzealousness in completing surveys, which led him to disobey the captain's orders. He became a surveyor in South Carolina and found that the biprovincial commission to survey the North Carolina–South Carolina boundary line had run the line eleven miles too far south, as far west as the Catawba River. As a result of Cook's astronomical observations, the line was run, west of the Catawba, twenty-two miles north in compensation, as can be seen on Cook's and all subsequent maps of North Carolina and South Carolina.

In the Naval Library in London are two manuscript maps of parts of the Gulf coast made by James Cook.[26] They are in two folio guard books that contain a unique collection of eighteenth-century manuscript charts of surveys of the Atlantic and Gulf coasts of North America. In addition to the Cook maps and to the Cyprian Southack manuscript, mentioned earlier, the volumes contain over one hundred other charts.

The extraordinary achievements of British hydrography in

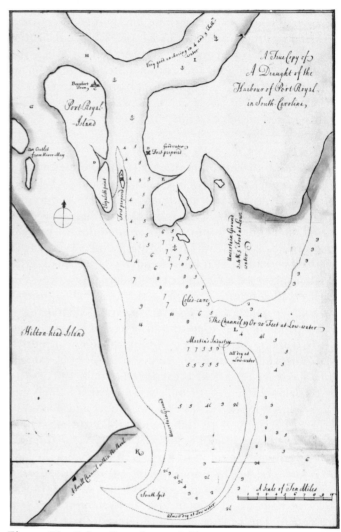

Figure 22 John Gascoigne, *A True Copy of A Draught of the Harbour of Port Royal . . .* (1729) [manuscript]. Crown-copyright reserved. Courtesy of the Public Record Office, London (Adm. 1/1827).

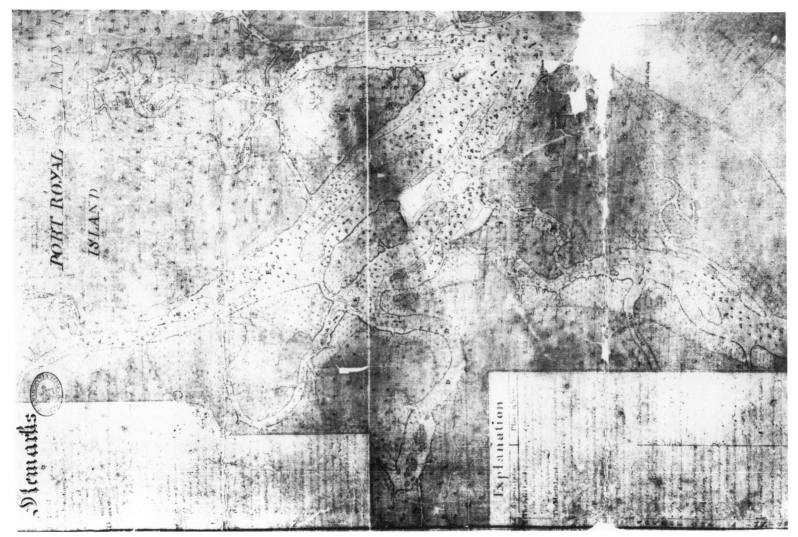

Figure 23 Francis Swaine [detail] from *Port Royal* (1729) [manu- Office, London (C.O. 700/Carolina 7).
script]. Crown-copyright reserved. Courtesy of the Public Record

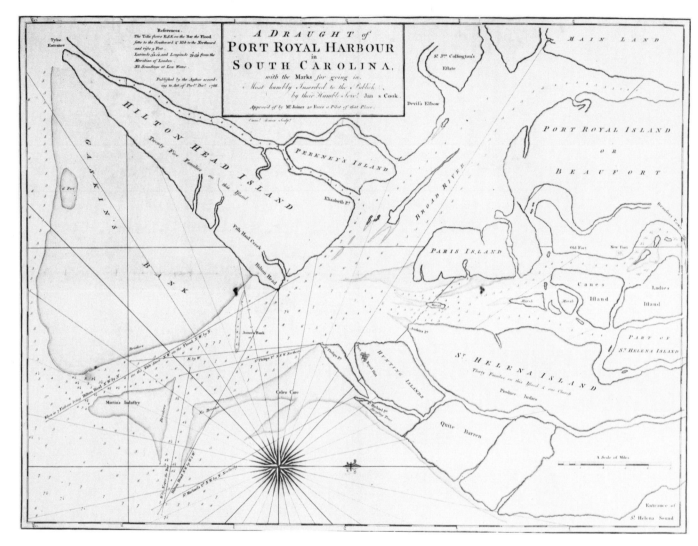

Figure 24 James Cook, *A Draught of Port Royal Harbour* (London, 1766) [separately published]. Courtesy of the William L. Clements Library.

North American waters between 1760 and the end of the Revolution are accounted for chiefly by the men who made them, but also by the developments in method and instruments used that characterized this period. The usual method of preparing charts before the middle of the eighteenth century is well described by Murdock Mackenzie, who in 1744 began to make the first accurate charts in the British Isles (the Orkneys) based on triangulation, with careful measurement of a base line on land and the use of a theodolite to measure angles. The usual way of taking a marine survey, he said, is

sailing along the land, taking the bearings of the headlands by a sea compass, and guessing at the distance by eye or log line; and also from the common tho' less certain method of constructing charts, either without surveying, navigating, or viewing the plans themselves; but only from verbal information, copied journals, or superficial sketches of sailors casually passing along the coast.

More careful surveying of a coastline, as Mackenzie said, involved the use of a ship's log: that is, a piece of wood dropped from the stern of a ship. The log (supposedly) remained stationary, attached to a knotted rope that was run out from a reel. At the end of a quarter or half minute by the running glass, the rope was pulled in and the number of knots were counted to calculate the speed of the ship. In marine terminology, therefore, the knot came to mean a unit of speed equal to one nautical mile per hour. Variations in the strength of the wind and the flow of currents obviously made the estimates inexact; but even this method, as Mackenzie indicated, was not always used.

Greater use of triangulation in surveys; employment of improved theodolites[27] for measuring angles from and along the shore; greater accuracy in recording latitudes; more frequent notation of compass variation;[28] and marked increase in the number of soundings along the coast and in harbors: these were some of the advances in method. Another important advance was in ease and accuracy of determining longitude. At the beginning of the eighteenth century distances across the Atlantic on charts were often wrong by many degrees; navigators were wrecked on shores that they thought hundreds of miles away. As late as 1753 the Marquis de Chabert, who had made longitudinal observations off the Newfoundland and Canadian coasts, wrote that current charts were so inaccurate that in some cases they showed variations of 9° of longitude ("neuf degrés de longitude" = 540 nautical miles) for the same position.[29] The publication in 1767 of the *Nautical Almanac*, in use with Hadley's reflecting octant (1731),[30] a forerunner of the sextant, made possible accurate determination of longitude by computing lunar distances.[31] The mathematical computations were laborious; but John Harrison's chronometer,[32] invented and tested by 1761, was still too expensive an instrument for practical use. Captain James Cook did not have one until his second voyage of discovery in the Pacific. Above all, however, the advance that occurred was caused by the appearance of a highly qualified group of men, devoted to their mission and supported by interested government agencies.

DE BRAHM, HOLLAND, COOK, AND DES BARRES

The French and Indian War had made clear the necessity of charting the islands, shoals, harbors, and coasts from Canada to the West Indies. No department of the government was authorized to undertake such a survey; the Hydrographic Office was not established until 1795. The program that was to culminate in the great compilation of charts of the American coast called *The Atlantic Neptune* began, therefore, under a number of different auspices, sometimes with different purposes and different levels of performance. At first there was no general plan for publishing charts; at the beginning Thomas Jefferys and other private publishers produced many of them. The governor or assembly of a province supported some of the surveys; the Board of Trade and Plantations ordered others; and the Admiralty, with its resources, undertook still more.

In the South, William Gerard De Brahm, of whose land mapping I spoke in the first essay, took with him to London in 1772 charts and a manuscript that he published as *The Atlantic Pilot*.

Its charts of Florida were poorer than much of his other work, such as the great twenty-by-five-foot map of the peninsula now in the Public Record Office;[33] but his *Atlantic Pilot* is a pioneer attempt to examine the source and nature of the Gulf Stream. His former deputy surveyor general, Bernard Romans, published in 1774 a map of the coast of the Florida peninsula that, though not the equal of *The Atlantic Neptune* in esthetic qualities, was the first large-scale, detailed, and accurate printed survey of the peninsula and the adjacent west coast. As the Revolution approached, Romans, who was living in Hartford, Connecticut, cast his lot with the Americans. In April 1775 he was a member of the commission appointed to take Ticonderoga and its outposts. Disagreeing with the mode of attack, he went off by himself and captured Fort George single-handed. This was not as impressive a feat as it sounds; there was only one invalided officer and his assistant in the fort. The officer said Romans treated him "very genteel and civil."[34] Romans was later taken prisoner and died (apparently) on shipboard on his way back from the West Indies in 1784.

Samuel Holland, already referred to in the second essay for his part in Governor Murray's great survey of Quebec, had been a lieutenant in the Dutch artillery before going to England about 1754. Two years later he arrived in America as a lieutenant in the 60th Regiment of Foot. This was a unit of the Royal Americans, recruited from Americans, with a number of officers who had a European background and professional training. Holland's rise was rapid. A military engineer and a fine draftsman, he was ambitious and a hard worker. In 1764, Holland, by that time a captain, was appointed by the king to be surveyor general of the Northern District of North America. His first instructions were to survey Prince Edward Island, the Magdalen Islands, and Cape Breton Island. These surveys were used for *The Atlantic Neptune*. He then moved on to New England, and was the surveyor for many *Atlantic Neptune* charts of this area. T. Wheeler and J. Grant, two of Holland's assistants, made a chart of Boston Harbor in 1775.[35] Before the Revolution Holland had completed his surveys of the New England coast and had moved on to New York; his head-quarters were in New Jersey when the war forced him to leave.

Many years before the Revolution, the day after the fall of the French stronghold of Louisbourg in 1758, the young sailing master of Captain Simcoe's ship *Pembroke* saw Holland surveying Kennington Cove, and asked him for instruction in surveying techniques. The young warrant officer became more famous than his instructor. He was the most famous of the James Cooks, a self-educated native of Whitby, Yorkshire, who had grown up in poverty, gone to sea, and was already beginning his rapid rise, phenomenal in the Royal Navy at that time, to positions of trust and importance. Before the end of the year Cook had surveyed and drafted a chart of Gaspé and Chaleur bays that so impressed his captain, Simcoe, that it was forwarded to London for immediate publication. This was Cook's first chart, to be followed by many of all parts of the world. Cook's first important assignment was the charting of Newfoundland. These charts, together with others by Cook's successor, Michael Lane, were published by Jefferys in his *Newfoundland Pilot* (ca. 1769).[36] Des Barres also published in *The Atlantic Neptune* many of Cook's charts, which are notable for their accuracy and skill in depicting a very difficult coastline, heavily indented, with many islands and hazards. A number of Cook's manuscript charts are now in the Hydrographic Department in Taunton, England. "Old Ferrol Harbour, Newfoundland," made in 1764 (fig. 25),[37] is one of the very few charts that show how Cook worked, since on it he has drawn, in dotted lines, the angles he measured to form his triangulation, and in places he has penciled very lightly the distances from one point to another. The style of the chart is typical of all Cook's work; it is delicately colored in browns and greens.

Joseph Frederick Wallet Des Barres's chief claim to fame is *The Atlantic Neptune;* however much others contributed to its success, his was the intellectual acumen, the driving physical stamina that brought it to a successful completion. He was born in Switzerland in 1721 and died in Halifax, Nova Scotia, on 24 October 1824, a month before his one hundred and third birthday. It was his good fortune to begin his mathematical training at the University

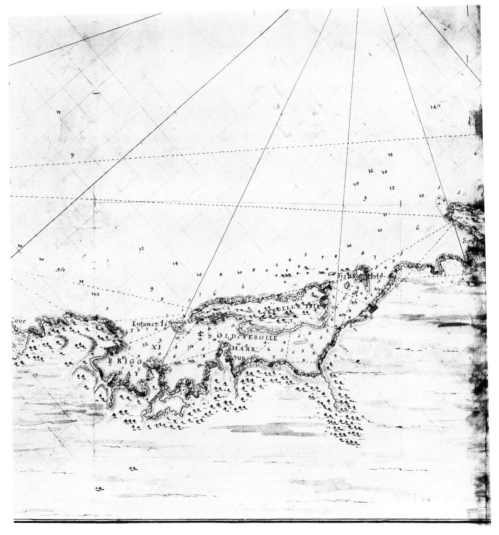

Figure 25　James Cook [detail] from *Old Ferrol Harbour, New-foundland* (1764) [manuscript]. Courtesy of the Hydrographic Office, Taunton.

(6) *Chebucto Head North ¼° West distant 2 Miles*

Surveyd, & Published according to Act of Parliament by J.F.W. Des Barres Esqr. Janʳ 30. 1779.

Figure 26 J. F. W. Des Barres [detail] from *Chebucto Head . . .* of the Ayer Collection, the Newberry Library.
(1779), in: *The Atlantic Neptune* (London, 1774–1800?). Courtesy

54

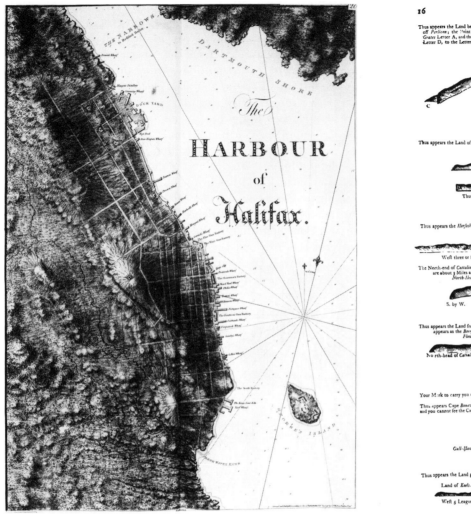

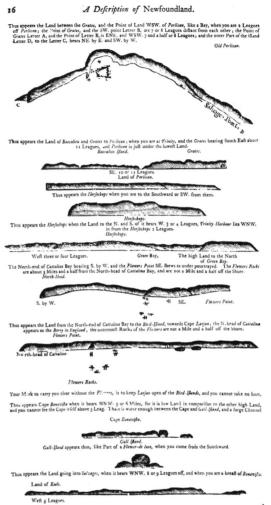

Figure 27 J. F. W. Des Barres, *Halifax Harbour* (1779), in: *The Atlantic Neptune* (London, 1774–1800?). Courtesy of the Ayer Collection, the Newberry Library.

Figure 28 [Coastal profiles] from *The English Pilot: The Fourth Book* (London, 1783). Courtesy of the Ayer Collection, the Newberry Library.

of Basel under Daniel Bernouilli, who has been called the founder of mathematical physics, and his brother James, who contributed to the practical application of the calculus. He became a British subject and trained as a military engineer at the Royal Military College at Woolwich; he was appointed lieutenant in the Royal American Regiment in 1756. At the siege of Louisbourg under General Amherst he may have met two other officers who played an important part in the making of *The Atlantic Neptune,* Holland and Cook. At Quebec he is said to have been making a report to General James Wolfe when Wolfe was mortally wounded, but one historian of that battle has said that if all the persons reported to have been by Wolfe at his death were present it is a wonder that anyone was left to fight the French!

From 1763 to 1774, with headquarters in Halifax, Des Barres surveyed the coasts of Nova Scotia. In 1774 he went to England and spent the next ten years preparing his charts and those of other military and naval officers. At first he had half-a-dozen assistants; but with the outbreak of the Revolution the need of charts for British fleets became urgent. From 1776 to 1779 Des Barres crowded two houses with twenty assistants. The total number of plates produced under Des Barres's supervision is not known; no adequate bibliographical study of the subject has appeared. Des Barres himself referred to 257 plates; but I. N. P. Stokes states that the New York Public Library has upward of thirty not mentioned by Des Barres, possibly because he discarded those charts and views that did not come up to his standards.[38] The Library of Congress has slightly fewer than 250; but it has 2,300 charts and views that have not been collated. Of some charts in *The Atlantic Neptune* there are ten or more variants or issues. Des Barres divided the work into five books: Nova Scotia; New England, which Holland surveyed with his assistants, chiefly Thomas Hurd, later hydrog-

rapher of the Navy, Blaskowitz, Wheeler, and Grant; the Gulf of Saint Lawrence, where Cook made most of his contributions, especially in the charting of Newfoundland; the coast south of New York; and various views of the American coast. There is, however, no copy of *The Atlantic Neptune* that is complete; in fact, no two sets are alike, for the plates were selected for a particular purpose or tour of duty.

Although Des Barres was contemptuous of land surveying in comparison to sea charting, his maps show differences in soil structure and types of beaches, and often include details of the fields, woods, and houses adjacent to the shore. His plans of harbors and the number of soundings in them, especially of those he surveyed himself, are impressive in contrast to any previous work (fig. 26). But it is the artistic excellence of his views, the sheer beauty of the work on them that he did or supervised, that gives a pleasure only those who open the pages of *The Atlantic Neptune* can realize. Contrast the engravings of the profiles of shoreline in an *English Pilot* or a *West-India Pilot* (figs. 27–28). In a time when there were no photographs, when the seaman could not tell his latitude accurately and often could not count on his charts to be accurate, the identifying profiles and directions accompanying the charts were of vital importance. Des Barres, in such views as are shown in the figures here reproduced, raised the drawing of coastal profiles to a high art.

Des Barres's achievement as a hydrographer was immediately and widely recognized. In Paris, *L'Esprit des Journaux,* in reviewing *The Atlantic Neptune,* said: "[It is] one of the most remarkable products of human industry that has ever been given to the world through the arts of printing and engraving . . . the most splendid collection of charts, plans and views ever published."[39] Two hundred years later, the judgment has not changed.

The Cartography of Conflict

Map-making Related to the
French and Indian War and the Revolution

This essay brings us to the parting of the ways between England and America. It is interesting to a cartophile to note how many of the reasons for the violent rupture were geographical. When England began successfully to colonize North America in the seventeenth century, she thought she was settling a narrow strip of land, a sort of middle island en route to Asia. One has only to look at the Farrer map (fig. 1) to remind oneself of this. She had no idea that what she had was an enormous continent with all kinds of potentialities. By the end of the eighteenth century, she had found this out, and so had the American colonists. Thomas Paine used this argument with great force in a powerful propaganda pamphlet, *Common Sense,* often listed as one of the sparks that ignited the American Revolution. "Small islands, not capable of protecting themselves," he wrote,

are the proper objects for kingdoms to take under their care; but there is something absurd in supposing a Continent to be perpetually governed by an island. In no instance hath nature made the satellite larger than its primary planet; and as England and America, with respect to each other, reverse the common order of nature, it is evident that they belong to different systems. England to Europe; America to itself.[1]

In these essays we have seen maps arising to meet various demands; the need for knowledge of unexplored territory and the advancing frontier; the need to record the location of settlements and settlers, to solve boundary disputes and land claims, to show routes and roads, to make water routes and landings passable and safe; and last, the tremendous need of the parent country, far away, to know something about what this bear-like continent was that it had by the tail, to try to plan for its governing, development, and use.

Now was added another tremendous exigency: the fighting of two wars, at a great distance, on unfamiliar and difficult terrain, to determine whether France or England should control eastern North America, and, later, to determine whether or not England should retain her transatlantic colonies. There had been conflict all along, and it had stimulated mapmaking: with the Indians everywhere, with the Spanish in the debatable land between Florida and Carolina, with the Dutch in New York, and with the French. But this was out-and-out war, over a greater terrain and between great powers.

What kinds of maps does war need? Charts of harbors and rivers to put ships there safely, to blockade or avoid blockade; the sea was then, as in many another war, the British lifeline. Maps of terrain in which campaigns must be waged, which vary from Lord Percy's pen-and-ink sketch of roads from Menotomy to Cambridge to Jeffrey Amherst's maps of the wilds of northern New York. There are quickly made maps of campaigns and of battles to be fought at some small outpost, careful maps to keep account of what has happened and where troops are, maps that are reports of officers to their superiors, diagrams of enemy forts to be taken or of new forts to be built. And, finally, there are commercial maps to keep the public informed.

What sort of person makes the maps of war? First the frontiersmen: pen-and-ink drawings based on reconnaissance surveys and crude sketches by frontier scouts were often the best delineation of the back country available for military leaders. Then follow the professional military engineers, and also many clever young officers

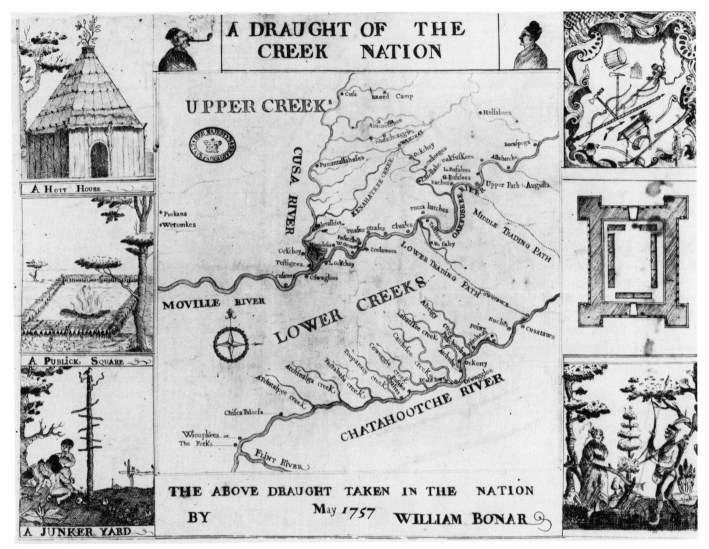

Figure 29 William Bonar, *A Draught of the Creek Nation* (1757) [manuscript]. Crown-copyright reserved. Courtesy of the Public Record Office, London (C.O. 700/Carolina 21).

drafted for the purpose. After the middle of the eighteenth century, the standards became amazingly good. There were Des Barres's beautiful profiles of the coast, but there were also fine sketches of the interior that accompanied military maps.

Consider William Bonar's 1757 map of the Creek country in Alabama, embellished with vignettes of Indian life: A Host House, a Public Square, a Junker Yard, Indian weapons, and an unusual profile of an Indian smoking a pipe (fig. 29).[2] Bonar, disguised as a pack-horseman, penetrated Fort Toulouse and brought back valuable military information; he also played a part in winning Creek tribes away from their French alliance. Two years after making his map, and after a dramatic rescue by Creek allies from French captivity, he was in command of Fort Johnston as a lieutenant in the provincial militia.[3]

The two engagements between General William Johnson and the French General Dieskau, on 8 September 1755, are vividly shown on Samuel Blodget's "Prospective plan of the battle near Lake George . . . ," of which a slightly smaller copy was engraved by Thomas Jefferys and published the following year, 2 February 1756, with minor changes in the placement on the sheet of the map of the Hudson River and of the insets of Forts Edward and William Henry.[4] The plate shows two views: to the left, the ambuscade in which Ephraim Williams, the English colonel, and Hendrick, the Mohawk chief, were killed; and to the right, the abortive attack of Dieskau on the hastily formed barricade of wagons and tree-trunks of Sir William Johnson's forces. Baron Dieskau was captured. It was a heartening British victory after Braddock's defeat. Blodget, a sutler in the Massachusetts Rangers, was himself an eye witness of the battle; the "Prospective plan," engraved by Thomas Johnston in Boston by 22 December 1755, is the earliest engraving in the British colonies to show an American historic scene. In his accompanying pamphlet of explanations with a carefully numbered key, Blodget comments that "Both the Canadians and Indians became invisible to our men, by squatting below the undergrowth of shrubs, and Brakes, or by concealing themselves behind the Trees." Jefferys states more concisely that

his "Prospective View" is "the only Piece that exhibits the American method of Bush Fighting."

A superb military chart, differing in technique and quality from the preceding maps and drawn at the end of the Revolutionary War period by an expert surveyor and draftsman, is "A Plan of the Posts of York and Gloucester in the Province of Virginia . . . Surveyed by Captn. Fage of the Royal Artillery."[5] It shows the operations of the opposing armies that terminated in the surrender by Cornwallis at Yorktown on 17 October 1781. Hachuring indicates the elevations and depressions; ravines, fields, woodland cover, and marshes are shown. Besides the location and movements of the ground forces, the chart shows the position of ships in the York River, including the British ships fired and sunk by a French battery. An overlay sheet, usually but not always attached to the bottom left corner of the main map, shows the outer defenses southeast of Yorktown constructed by Cornwallis in September, from which he had to withdraw because of heavy attacks by French and American troops: "The Position of the Army between the Ravines on the 28th. and 29th. of Sept. 1781." Captain Edward Fage's survey was engraved and published in 1782 for *The Atlantic Neptune*; it is a fine example of Des Barres's professional cartographic skill.

RISING TENSIONS AND THEIR CARTOGRAPHICAL EXPRESSION

Before open war there was a long period of increasing friction as it became obvious that British and French interests in America were going to conflict, especially in the areas of fishing and commerce along the northern coast and in the control of the valuable Indian fur trade and the advancing British frontier in the trans-Allegheny region. After 1750 the rising tensions led to many pamphlets vigorously supporting British claims and attacking the French; often accompanying them, or published independently, were maps showing the boundary lines supported by each side. Thomas Jefferys, as has been mentioned before, was a vigorous Francophobe. In 1750 he published a map of Halifax, Nova Scotia, established the preceding year to balance the mighty French fortress of Louisbourg.[6] In New England especially, the establishment of Halifax caused exultation. In 1755 Braddock Mead, Jeffery's geographer, drew a fine map of Nova Scotia, accompanying it with a pamphlet that attacked the French.[7] Jefferys followed this up the same year by reprinting two smaller copies of a French map of Cape Breton Island and Nova Scotia, one above the other; the upper showed French claims, the lower the English.[8] Likewise, in 1755, a large map of the British Empire in North America was put out by a group who called themselves "A Society of Anti-Gallicans."[9] It showed French encroachments by a brown boundary line and the "just and rightful" English claims in purple. The English line runs up the Mississippi to Lake Michigan and thence to Lake Abilibis near Hudson Bay. The English already controlled Hudson Bay; the "anti-Gallicans" thus, "justly and rightfully," took possession of all North America east of the Mississippi except for a restricted area from Lake Huron eastward and along the Saint Lawrence River valley and gulf. Numerous other maps of North America showing British "rights" and French "encroachments" were published in England in 1755, though they were not as blatantly aggressive as the "anti-Gallican" map. Thomas Bowen and J. Gibson produced a large wall map, "An Accurate Map of North America . . . Exhibiting the Present seat of War and the French Encroachments." Huske's "A New and Accurate map of North America (where in the Errors of all preceding . . . maps respecting the rights of Great Britain, France & Spain . . . are Corrected)" accompanied *The Present State of North America* (London and Boston, 1755) and is found in other contemporary works.[10]

THE FRENCH AND INDIAN WAR

There had been no real peace along the frontier for some time before the middle of the century. Now came open and declared war, the North American part of the general struggle known in Europe as the Seven Years' War. The nature of the fighting dictated the nature of the maps needed; it was both a frontier war and, in its later development, a slow, cautious advance along lim-

ited routes that involved the construction of forts.[11] The American conflict may be summarized as the effort of the French to contain the English colonies south of Cape Breton Island and east of the Appalachians, asserting their supremacy over Canadian waters and the valley of the Ohio, which joined French Canada with the French possessions down the Mississippi to New Orleans. England's part in the war was to break out of this encirclement. The French had added to their northern fortifications, such as Louisbourg on Cape Breton Island and Forts Saint Frederick and Ticonderoga in the Lake Champlain region, a new circle of forts from Presqu'Isle on Lake Ontario to Fort Duquesne at the forks of the Ohio where Pittsburgh now stands. These the English had to reach through poorly mapped territory.

On 31 October 1753, George Washington made a journey that initiated far-reaching consequences when he started for Fort Le Boeuf, built near the present Waterford, Pennsylvania, with a letter from Governor Robert Dinwiddie of Virginia to the French commander, asserting that the trans-Allegheny region of the Ohio Valley is "notoriously known to be the Property of the Crown of *Great-Britain*."[12] Legardeur de Saint Piere's polite but firm letter of refusal to abandon his post Washington carried back to Williamsburg, which he reached on 16 January 1754. His own manuscript map, which traces his trip from Fort Cumberland on Will's Creek, a branch of the Potomac River in Maryland, to Fort Le Boeuf, was sent to London by Governor Dinwiddie; Washington had pioneered a route, later used by Braddock, that became a major highway between western Maryland and the forks of the Ohio.[13]

As guide, Washington had taken with him Christopher Gist. Few if any English frontiersmen were at that time more knowledgeable than Gist concerning the Ohio Valley; he had explored northeastern Kentucky in 1750–52, eighteen years before Daniel Boone's better known expeditions, and in 1753 had established a plantation near the Youghiogheny River. When, in July 1755, Major-General Edward Braddock approached Fort Duquesne, Gist went ahead to reconnoiter; his "The Draught of Genl. Braddocks Route towards Fort Du Quesne as deliver'd to Capt. Mc-Keller Engineer. By Christ. Gist The 15th of Sept. 1755" gives not only the details of the march from Will's Creek but also shows a knowledge of the surrounding terrain and river systems.[14]

A series of six contemporary maps showing the order of Braddock's march to Fort Duquesne exhibits competent draftsmanship and was made by one of the accompanying British officers, Braddock's aid-de-camp Captain William Orme.[15] They give in detail the disposition of the army, with sappers ahead clearing a broad roadway, regiments in close order as if on parade, followed by artillery and supply wagons. If they show the ability of a British army to build a military road through the American wilderness, they also exhibit graphically the ignorance of or contempt for frontier warfare that led to Braddock's defeat. The British were in the open, unprotected and in battle order, while the French and Indians ambushed them from the surrounding forest cover.

For almost three years after the French victory at Fort Duquesne and Braddock's death from wounds in the fight came a series of incompetent military commanders, Governor William Shirley of Massachusetts, Daniel Webb, Lord Loudoun, and James Abercrombie, accompanied by inadequate support by the London and provincial governments. Except for the defeat of General Dieskau by General William Johnson at Lake George (1755) and the pitiful subjugation and transportation of the Acadians by Colonel Robert Monckton and Lieutenant-Colonel John Winslow in their Nova Scotian expedition, the record for the British was almost uniformly bleak. During this same period, however, a number of officers, engineers, and draftsmen were sent to America with the British troops whose skill and achievements in surveying, in plans for needed blockhouses and forts, and in mapmaking became increasingly apparent during the war and in the years that followed. Among these were the nineteen-year-old ensign John Montresor, wounded at Monongahela while serving as General Braddock's assistant engineer in building the road to Fort Duquesne; Bernard Ratzer, commissioned lieutenant in 1756; Captain Thomas Sowers, engineer of the fortifications at Albany and later at Oswego and Niagara; Captain Thomas Abercrombie, whose sketch of Lake George in 1756 was followed by a series of plans of forts along the route to Montreal; Harry Gordon, who in 1766 became chief engineer in North America; William Brasier (Brassier, Brazier), a skilled draftsman for Captain Sowers and others as well as surveyor and mapmaker himself; these and others often known only by the signatures to their work and a brief note in the Army lists produced the maps and plans of the period. A number of them continued their work into the Revolutionary years.

The immediate need felt by the colonies and by Governor Shirley, who succeeded to command after Braddock's death, was a line of perimeter defenses. "A sketch of . . . forts lately built on the frontiers . . . and their scituation with respect to the french forts on the Ohio & Lake Erie"[16] was drawn by William Alexander in 1756 and shows the determination of General Shirley to keep the French at least on the other side of the Alleghenies. Alexander, Shirley's secretary, though unsuccessful in winning his claim to an earldom in 1757, was later known as Lord Stirling and became a general in the American army during the Revolution.

With the arrival in Canada in 1756 of General Montcalm, the ablest general so far on the American scene, bringing French reinforcements, a series of further British defeats ensued, with the capture of several forts in northern New York province. General James Abercromby (Abercrombie), described as "an aged gentleman, infirm in body and mind" and "Booby-in-chief" by those under him,[17] succeeded through incompetence and bad judgment in having his fifteen thousand men defeated by Montcalm, with scarcely one-fourth that number, at Ticonderoga in the bloodiest battle of the war. Mante's "The Attack of Ticonderoga; Major General Abercromby, Commander in Chief," is a good plan of the battle.[18] Another Abercrombie, Captain Thomas, whose many excellent plans of forts include one of Ticonderoga,[19] drew a map of "the Scene of Action" in northern New York surrounded with eight insets of forts and surrounding areas, made in 1758 and dedicated to "Major General Abercrombie."[20]

After the dismal sequence of British defeats along the outposts and the slaughter of two thousand at Ticonderoga, the pendulum began to swing in the other direction. In 1757 William Pitt became secretary of war, began to weed out the incompetents, to appoint new commanders, and to give them adequate military support. To America he sent forty-year-old Major General Jeffrey Amherst, with a brilliant young man of thirty-one, James Wolfe, as second in command; their mission was to capture the French Gibraltar in America, the great fortress of Louisbourg on Cape Breton Island. With fourteen thousand troops and with unaccustomed coordination with the British navy under Admiral Boscawen, Amherst and Wolfe forced the capitulation of Louisbourg by the end of July 1758. Maps and plans of Louisbourg made before, during, and after its fall bear testimony to the careful collection of information for its siege; printed maps made by Jefferys, Rocque, and others reflect the triumphant reaction in England.[21]

Amherst could have followed up his victory at Louisbourg by pressing the attack against a demoralized enemy; instead, he avoided the disasters of his predecessors by a slow and careful multipronged advance up the Saint Lawrence River westward, by subduing the French fortifications at the Niagara River and Oswego to open the way from Lake Ontario northeast to Montreal, and northward along the Champlain valley. He moved his headquarters to Fort William Henry on Lake George in New York province, which he renamed Fort George.[22] Even the great victory at Quebec did not seem to hasten him; Wolf and Montcalm died on the Plains of Abraham on 13 September 1759, and Vice-Admiral Saunders effectively blocked further aid from reaching the French, but Amherst waited until the following year to attack Montreal.

The best printed map of the battle of Quebec is Thomas Jefferys's engraving, "A Correct Plan of the Environs of Quebec and of the Battle fought on the 13th. September, 1759: . . . Drawn from the Original Surveys taken by the Engineers of the Army." It is a beautiful production, showing the topography of the region in detail, with the location of the British ships and the lines, batteries, encampments, and attacks of the opposing forces (fig. 30).[23]

Although Amherst during 1759 engaged his own army in no decisive action, he did not allow his troops to remain idle. His own collection of maps shows that his forces were retaking lost forts and heavily engaged in rebuilding and strengthening old fortifications and building new ones. His engineers made surveys of strategic areas, executed drawings of existing defenses, and drew new plans for additional batteries, enlarged fortifications, and other forts in new locations. Some of these are sketches; others are carefully designed, colored, highly finished productions showing the competence of the makers and a pride in their work.

In July 1759 Amherst sent successful expeditions west to reoccupy Oswego and take Fort Niagara; of these captured forts Thomas Sowers, William Brasier, George Demler, Francis Pfister, and others drew plans.[24] In July and August 1759, Amherst moved in force to capture the French strongholds at Fort Carillon, which he renamed Ticonderoga, and Crown Point on Lake Champlain. Among the many maps resulting from this campaign, those by William Brasier may be noted. His "Plan of the Fortress and Dependent Forts at Crown Point . . . 1759" is probably the finest work in design, coloring, and detail of a very fine draftsman;[25] and his "A Survey of Lake Champlain, including Lake George, Crown Point and St. John . . . By William Brassier, Draughtsman, 1762," was engraved in 1776 for use by the army during the Revolution.[26]

In 1760 Amherst's plans matured; on 8 September Governor Vaudeuil surrendered Canada to Amherst at Montreal, surrounded by ten thousand troops that had sailed with Amherst from Oswego, the forces under Brigadier James Murray from Quebec, and the three thousand troops under Brigadier Frederick Haldimand from Lake Champlain. Although the French had capitulated, fighting continued; the Indians on the frontier kept up their hostilities, resentful of continued English expansion into their hunting grounds.

In the south, although De Brahm had made plans for extensive fortifications and a number had been built in anticipation of French attacks that never came, it was the construction of Fort

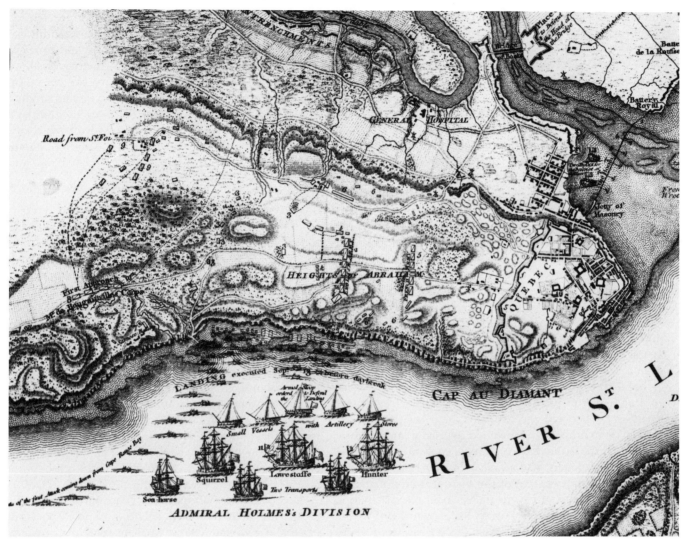

Figure 30 Thomas Jefferys [detail] from *A Correct Plan of the Environs of Quebec . . . 1759,* in: Thomas Jefferys, *A General Topography of North America* (London, 1768). Courtesy of the General Collection, the Newberry Library.

Loudoun on the Little Tennessee, with De Brahm as engineer in charge, and the serious aggravations perpetrated by Governor William Henry Lyttelton of South Carolina that caused the outbreak of the Cherokee War.[27] Three interesting maps resulted from the war and its aftermath. Captain John Stuart, former trader and future superintendent of Indian Affairs in the Southern Department, who escaped in the massacre after the fall of Fort Loudoun in August 1760 with the aid of a friendly Indian chief, drew a map showing "The Road by which Capt. Stuart escaped to Virginia" and delineating the Appalachian mountain range between northern Alabama and Virginia.[28] Made about 1761, it is a forerunner of the series of great wall charts that Stuart executed or supervised over a dozen years later. After hearing of the fall of Fort Loudoun, General Amherst sent Lieutenant-Colonel James Grant, later to become governor of East Florida, with troops from New York. Grant reached Charleston early in 1761; before midsummer the Indians, many of their villages laid waste, sued for peace. A large map of Grant's expedition shows his route and the surrounding country; it is an unsigned sketch made for Amherst shortly after the campaign was finished.[29] After the cease-fire of 19 November, Lieutenant Henry Timberlake, who had seen service under George Washington, volunteered to go on a peace mission from Virginia. He made copious notes of the country he traversed and later published "A Draught of the Cherokee Country" with details of Indian paths and the location of their settlements.[30]

During the years between the two wars, many of the military engineers and their draftsmen continued to survey and produce maps. Bernard Ratzer, who was commissioned lieutenant of the Royal American Regiment of 60th Foot, on 20 February 1756, eventually became senior captain of the First Battalion, outranking Des Barres.[31] His later surveying and cartographical activities, however, extended beyond strict military assignments. In 1766–67 he made a survey of New York City, which was the basis for Faden's magnificent map of 1776. In 1769 he surveyed the boundary line for the commissioners of New Jersey and New York in the long-disputed division between those two provinces. In Jamaica, to which the First Battalion of the 60th Foot, or elements of it, was moved in 1773, he was employed by the assembly to survey the acreage of the island as well as its forts and fortifications.[32] He apparently did not return to North America during the Revolution. Samuel Holland, who became surveyor-general of the Northern District in 1764, made extensive surveys and produced or supervised the making of many maps that were used in the Revolution. John Montresor, whose activities extended from the beginning of the French and Indian War to the Revolution, became chief engineer of the army in North America under Howe and Clinton.[33]

THE REVOLUTIONARY WAR

The cartography of the Revolution has some marked differences from that produced by the French and Indian War. The majority of maps in the earlier conflict are small-area plans of fortifications in upper New York and in the maritime provinces, although alarums and excursions along the frontier and in the south resulted in markedly increased geographical knowledge of the interior. The Revolution produced a great body of plans of forts and local topography, but it also resulted in many maps portraying engagements and battles, delineating movement of troops, and showing details of campaigns conducted over extensive areas.[34]

At the beginning of the Revolution or shortly before, a number of important regional printed maps appeared, often based on those available twenty years before but incorporating greatly increased topographical information such as roads helpful for military planning. These include Montresor's *New York* (1775, 1777) (fig. 31),[35] Sauthier and Ratzer's *New York* (1776),[36] Scull's *Pennsylvania* (1770),[37] and Mouzon's *North and South Carolina* (1775).[38]

After 1775 came a rapid resurgence in the production of maps mostly for immediate military needs or for reports submitted by officers in command. Most of them remain in manuscript; since there was no way to duplicate maps mechanically except by copperplate engraving, several copies, often with additions or omissions, were made of military maps when needed. These hand copies, made and signed by the officer himself or at times copied by a civilian draftsman attached to a command,[39] explain the presence of similar manuscript maps in different collections. Some copies are evidently drafted from originals made years earlier. Most are drawn, however, soon after the survey; some have the added note "taken on the spot" or "surveyed on the same day [as the battle]."

Although a few mapmakers survive from the earlier war, like Montresor and Holland, most of the names on revolutionary war maps are new ones.[40] These army officers and engineers show professional training. The Royal Military Academy for artillery and engineering cadets at Woolwich required training in drawing. The Corps of Engineers established a code of instruction that included scale regulations for different kinds of surveys and plans.[41]

THE WAR BEGINS: MASSACHUSETTS

The maps of the Revolution illustrate graphically the campaigns and phases into which the history of the war divides as its geographical center shifts. The primary movement in the war is from north to south, with divagations.

In the Percy collection in Alnwick Castle is a map that may be the earliest of the Revolution: "A Plan of the Town and Harbour of Boston. and the Country adjacent with the Road from Boston to Concord. Shewing the Place of the late Engagement between the Kings Troops & the Provincials, together with the several Encampments of both Armies in and about Boston. 19th. April 1775" (see title page).[42] The colored plan shows the location of the British and Provincial troops at various locations during the retreat from Concord, with some identifying legends. On the plan the marks across a stream showing the "Bridge where the attack began" is not Concord's "rude bridge that arched the flood" but another, over Mill Brook, where the British under Colonel Smith fired a parting volley and the watching militia closed in with heavy fire. Percy, whom General Gage had sent with reinforcements when he heard that Smith was in trouble, relieved Smith at Lexington, not west of it, as on this map; he took the lower road

Figure 31 John Montresor, *A Map of the Province of New York . . .* (1775), in: William Faden, *North American Atlas* (London, 1777). Courtesy of the Ayer Collection, the Newberry Library.

through Menotomy (Monatony on the map) toward Cambridge, rather than the upper fork, as the plan's author wrongly shows. The provincial army assembled around Boston with amazing speed; by 21 April about nine thousand surrounded the circle of campfires seen by the British from Beacon Hill. The placing of encampments on this map indicates that it was made at least by 26 April; on the 22d, many regiments were at Watertown, but were ordered to Cambridge on the 26th, where headquarters was established under General Artemas Ward. This copy of the map, belonging to Earl Percy, is the only one known; a John De Costa, however, otherwise unknown, must have obtained a similar one that he used as a basis for a more elaborate map, published in London at the end of July, on which he included the "hot" news of the Battle of Bunker Hill that had just reached England.

Closely related to the plan of the Concord-Boston area is an unsigned, untitled sketch of the country between Menotomy and Charlestown in the Percy collection, on two sheets pinned together, which shows the road from Menotomy to Cambridge as a straight line at the top.[43] On Percy's return from Lexington, he rightly sensed that he would run into an ambush ahead if he went through Cambridge as planned. At a tavern shown on this map, he wheeled his troops into "Kent's Lane through which the Troops return'd from Concord" (a road not elsewhere named, apparently, in the voluminous literature on that day's events) and marched toward Charlestown. The legends on the sketch map are fascinating: "Hilly, broken ground"; "each side lined with trees"; "up a long hill." There can be no better example than this map of the way in which ignorance or knowledge of a countryside and its roads can cause military disaster or escape. If Percy had not known of Kent's Lane and had marched on into the militia who had destroyed the bridge across the Charles River and were waiting in Cambridge, the British would have been decimated (fig. 32).

Another manuscript plan in the Percy collection shows the strategic peninsula of Charlestown, north of Boston, to which Percy retreated from Lexington on 19 April. It was drawn by Captain Edward Barron of the King's-own Regiment[44] and shows

the careful and high artistic standards characteristic of the military cartography of the period, with neat little crosses in the cemetery and the shadow thrown by every tree. Barron made it after the British, with heavy loss, had won the battle on 17 June and had built a strong fort on Bunker Hill, "ditched, palisaded and fraised"; fortified the Neck; and strengthened the "Rebel Redoubt" on Breed's Hill. Occupying regiments are represented by their colors. Percy was often here during the siege of Boston. He had been at the Roxbury defenses on 17 June, but his regiment was in the thick of the fight. The British landed on the north and west, climbing the slopes into withering fire from the redoubt and the famous fence, described by Captain Chester as ". . . a poor stone fence, two or three feet high, and very thin, so that the bullets came through."[45] The town, burned by shelling, was in ruins when this map was made, sometime between July 1775 and March 1776, when Barron sailed to Halifax with the evacuating British.[46]

In May 1776 General Howe, who had succeeded General Gage after the badly planned victory of Bunker Hill, exercised his troops "in line" at Halifax, as depicted in another drawing by Captain Barron. Forced to evacuate Boston by Washington's brilliant fortification of Dorchester Heights, and the foiling of Percy's counterattack by a "hurrycane," Howe took about ten thousand troops and eleven hundred Loyalists with their belongings to Halifax. In Barron's plan he is shown preparing for the battle for New York that culminated in the taking of Fort Washington. The regiment of Percy, now advanced in rank to a lieutenant-general, is shown at right center.

NEW YORK AND RHODE ISLAND

In late June of 1776 Howe and his forces landed on Staten Island; on 27 August he began a successful attack on the American forces on Long Island, details of which are shown on an engraved map by Major Samuel Holland.[47] Howe occupied lower Manhattan and after almost two months' delay forced Washington to retreat, in an engagement at White Plains on 28 October, which is shown on

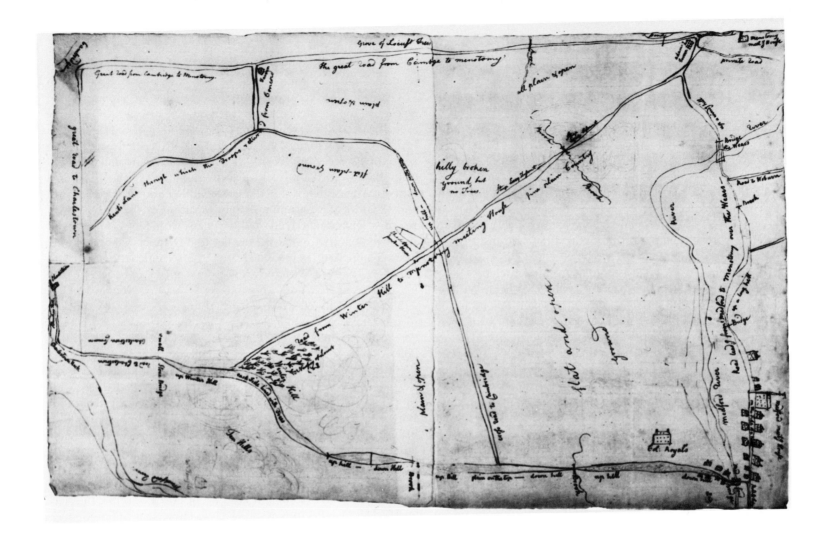

Figure 32 [A manuscript map of the country between Menotomy and Charlestown] (ca. 1774). Courtesy of the Duke of Northumberland, Alnwick.

Faden's engraving of a map by Sauthier. Howe attacked the American defenses on northern Manhattan Island in another successful engagement. Fort Washington fell on 16 November, and Percy ordered Sauthier to make a map of the battle. Sauthier began his surveys on the same day. Stokes calls the result "the most beautiful and probably the most accurate" revolutionary war map of northern New York Island (fig. 33).[48] It emphasizes Percy's part in the concerted attack of the British and Hessians on the fort; he forced his way from the south through the three lines of defense shown in the middle of the map, pushing the Americans back into the fort in his "ablest military action." Although the American loss here was heavy, Washington and the main army were already in New Jersey. In this map, Sauthier shows himself at his best as a surveyor and draftsman.

In December 1776, at the insistence of Admiral Richard Howe, who demanded a safe winter haven for his fleet, General William Howe sent Clinton and Percy to capture Newport. Rhode Island was soon taken, and Percy remained there as commanding general until 5 May 1777, when he returned to England after refusing what seemed to him Howe's unreasonable request that he send him troops from his own depleted forces. In the Percy collection in Alnwick Castle is a group of printed and manuscript maps of Rhode Island, made before, during, and after his occupation. They include Blaskowitz's fine "Topographical Chart of the Bay of Narraganset," dedicated to Earl Percy and engraved by Faden in July 1777.[49] It shows recent fortifications made by the Americans, but is largely based on surveys made under Holland's direction before the outbreak of the war. Two manuscript maps show the naval operation that on 22 May 1778 destroyed the shipyard and vessels on Fall River, a severe blow to the struggling American navy.[50]

THE HUDSON RIVER CAMPAIGN

The British plan for a campaign in 1777 to divide the American provinces at the Hudson and then to destroy effective colonial opposition in New England was a sound one. The governments on either side and the generals they appointed made a series of mistakes; those of the English were more serious and ended in the surrender of General John Burgoyne and his army at Saratoga on 17 October. Maps resulting from the campaign are numerous and give much topographical detail based on extensive surveying in the area of troop activity. Faden published "A Map of the Country in which the Army under Lt. General Bourgoyne acted in the Campaign of 1777 . . .";[51] "Position of the Detachment under Lieutt. Coll. Baum, at Walmscock near Bennington,"[52] the ill-advised foray in August that resulted in the complete rout of the Hessians; and "Plan of the Encampment and Position of the Army . . . at Bræmus Heights on Hudson's River,"[53] with the movements of Burgoyne's army from 20 September to 8 October, when he found himself hopelessly surrounded after the second battle of Bæmis Heights. Burgoyne with his army from Canada and Clinton with his army from New York were to meet at Albany. Neither reached Albany; but Clinton's forces reached and captured two strategic forts on the Hudson on 6 October 1777, the day before Burgoyne's defeat. Faden published a "Plan of the Attack of the Forts Clinton [West Point] & Montgomery. . . ," made by John Hills, Clinton's surveyor.[54] In 1779 the outposts of West Point at Stony Point and Verplanck's Point, which had changed hands several times, were captured from the British in a surprise attack by General Anthony Wayne. This action and the surrounding area are shown in a beautifully designed map by Hills, "A Plan of the Surprise of Stoney Point," finely engraved by Faden (fig. 34).[55]

THE NEW JERSEY–PENNSYLVANIA CAMPAIGN

After the fall of Fort Washington and of Fort Lee across the river on Harlem Heights in November 1776, Howe sent Major General Earl Cornwallis in pursuit of Washington, who retreated across New Jersey and the Delaware River. Recrossing the river on the night of 25 December, Washington surprised and routed the Hessians at Trenton and three regiments of British at Princeton on 3 January 1777.[56] Howe withdrew his troops to east New Jersey, and Washington wintered at Morristown.

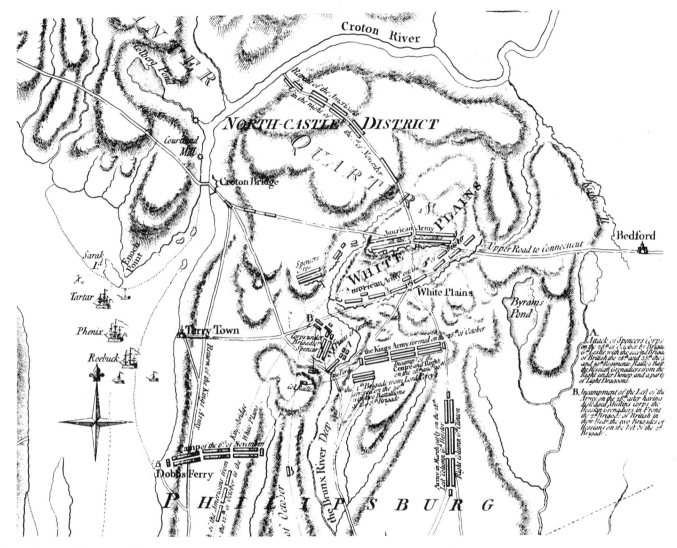

Figure 33 C. J. Sauthier [detail] from *A plan of the operations of the King's army . . . in New York . . .* (1777), in: William Faden, *North American Atlas* (London, 1777). Courtesy of the Ayer Collection, the Newberry Library.

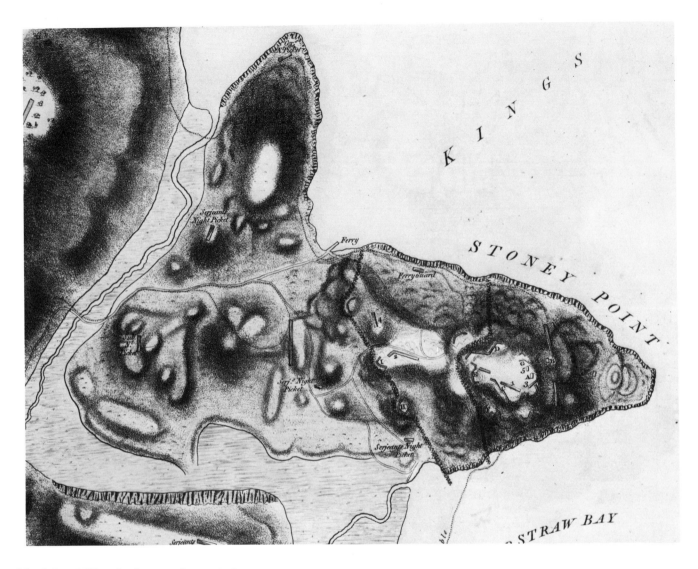

Figure 34 John Hills, *A Plan of the Surprise of Stoney Point*
(London, 1784) [separately published]. Courtesy of the General
Collection, the Newberry Library.

70

In August 1777 Howe sailed from New York with nearly nineteen thousand troops to the head of Chesapeake Bay; from there he marched north to Philadelphia. After a series of encounters, including the Battle of Germantown,[57] in which Washington was defeated, Howe spent the winter in Philadelphia, in complete control of the Delaware River valley, while Washington wintered in Valley Forge. In the spring Howe asked to be retired; General Sir Henry Clinton, who was appointed to succeed him, withdrew from Philadelphia in June 1778, pursued by Washington across New Jersey.

The best map of the New Jersey-Pennsylvania campaign is probably a large manuscript map in three sheets, now in the Percy collection, which extends from New York to Salisbury, Delaware, "Showing the Operations of the British Army Against the Rebels in North America from the 12th of August 1776 to the end of the Year 1779."[58] It shows opposing troop movements in color and is accompanied by a folio reference sheet with over eighty notations. Less elaborate but useful is Faden's engraving in 1784 of John Hills's "A Plan of Part of the Provinces Pennsylvania, and East & West New Jersey," which shows military operations from Howe's landing at Elk River in 1777 to Clinton's embarkation for New York at Navesink in 1778.[59] In the Library of Congress is a collection of twenty drawings by John Hills and others that records the operations of Clinton's forces in New Jersey and includes plans of New Jersey towns, counties, roads, and fortifications.[60]

THE SOUTHERN CAMPAIGN

Clinton, with a small naval force under Sir Peter Parker, had made an attack on Charleston in June 1776, hoping for a quick victory and wide support by Loyalists. Clinton, whose troops were on Long Island, South Carolina, could not reach Fort Sullivan because of a water barrier between the two islands; and gunfire from Fort Sullivan damaged Sir Peter's fleet.[61] The British returned to New York.

This unsuccessful expedition long rankled in Clinton's mind. The war in the North was not going well, but an attack on Savan-nah at the end of December 1778 by Lieutenant-Colonel Archibald Campbell was gratifyingly successful. Within six days of Campbell's landing with thirty-five hundred soldiers on 23 December, Savannah fell. General Augustine Prevost arrived in support with two thousand more troops from Saint Augustine; Augusta was captured on 29 January 1778, and Sir James Wright arrived from England to resume his duties as governor with local Loyalist backing. A counterattack on 4 October by Admiral d'Estaing with a strong fleet and four thousand French soldiers, supported by the continental general Benjamin Lincoln's troops and Count Casimir Pulaski's legion on land, was repulsed by General Prevost. D'Estaing sailed away. Several maps record these events, including two fine engravings by Faden.[62]

Encouraged by these successes in Georgia, Clinton decided to mount a campaign in South Carolina and there, as he thought, to attack the soft underbelly of the rebellion in the South by capturing Charleston. Lord George Germain, the colonial secretary in London, concurred. This time Clinton planned carefully and came with ten thousand men and with Admiral Mariot Arbuthnot's fleet in support. They left New York at the end of December; on 12 May 1780, General Lincoln, hemmed in on all sides, surrendered. The best map of the siege is Des Barres's large and handsome chart, "A Sketch of the Operations before Charlestown, the Capital of South Carolina. Published 17th of June 1780."[63] Another good map is Faden's "Plan of the Siege of Charlestown in South Carolina . . . 1787";[64] there are literally dozens of manuscript and printed contemporary maps of various stages in the siege, of the environs, and of troop entrenchments.

Clinton returned to New York, leaving Cornwallis in command with a premonitory warning not to undertake expensive and hazardous adventures. Cornwallis was not a defensive strategist; by nature he was aggressive. He soon had opportunity to show his energetic character. Faden's "Plan of the Battle Fought near Camden August 16th. 1780"[65] shows Cornwallis's attack against the superior force of General Gates near Saunder's Creek, which ended in the complete rout of the three thousand Continentals and militia. Gates returned to Hillsborough, North Carolina, with about seven hundred men.

A series of victories and reverses for Cornwallis, comparatively minor in themselves, created a situation that thwarted the triumphant march to Baltimore that Clinton had hoped for. Lieutenant-Colonel Banastre Tarleton, Cornwallis's feared and hated cavalry officer, has a map in his *History* that presents "The Marches of Lord Cornwallis in the Southern Provinces . . . Comprehending the two Carolinas with Virginia . . . By William Faden . . . 1787."[66] Tarleton himself lost heavily in reputation by a severe defeat at Cowpens, South Carolina, in January 1781 in an overconfident attack on General Daniel Morgan. General Nathanael Greene had replaced General Gates, and Cornwallis, determined to inflict a punishing blow, marched to Guilford, North Carolina. "Battle of Guildford Fought on the 15th. of March 1781," by Faden, shows the British winning the field, for Greene's forces retreated at the end of the day.[67] Cornwallis did not have the strength to pursue; he took his army to Wilmington, North Carolina, and Greene marched south to engage in a series of skirmishes that he mostly lost but that forced the British back to Charleston. "Sketch of the Battle of Hobkirks Hill, near Camden, on the 25th. April, 1781," a map with charming details drawn by Captain C. Vallancey of the Irish Volunteers,[68] shows an encounter, sometimes called the second Battle of Camden, in which Lord Rawdon, whom Cornwallis had left in command in South Carolina, gained the field but later retired to Charleston.

Meanwhile, in Wilmington, Cornwallis had time to do some thinking. This was an activity, one historian has remarked, in which he did not excel.[69] He decided to march north and take over command of British troops in Virginia; he reached Petersburg, Virginia, on 12 May 1781; an interesting manuscript map of Virginia shows Cornwallis's movements there, including expeditions he sent out from Richmond, until he reached Portsmouth.[70] A number of detailed plans of skirmishes made during this period were drawn by Lieutenant-Colonel J. G. Simcoe and officers in his Queen's Rangers. The forays themselves were usually not im-

portant; but the details on the maps, both of Virginia localities and of places in other colonies in earlier encounters, provide excellent topographical information not elsewhere duplicated.[71] Cornwallis, who had moved his troops to Yorktown and Gloucester, found himself hemmed in by superior military and naval forces. The maps and charts of the Yorktown campaign have been listed and described by Coolie Verner.[72] To his list may be added four manuscript plans of the naval encounter at the capes of Chesapeake Bay on 5 September.[73] They show the position of individual vessels of the French and British fleets at different stages in the Battle of the Capes. With the withdrawal of the fleet under Admiral Sir Thomas Graves and Sir Samuel Hood and its return to New York, Admiral François de Grasse reentered the bay. Cornwallis could escape neither by land nor sea. By mid-October he was without ammunition for his guns. He capitulated on 19 October, and the war virtually came to an end.[74]

The American Revolution produced a remarkable output of maps, exceptional in variety and quality. The great British engravers and publishers were responsible for the production and dissemination of many of these. No less fine are many of the manuscripts that were never printed; until recently only part of these have been known or available, even to scholars. Several great collections have been acquired by American libraries; the Percy and the Royal United Services Institution collections in England have been introduced in this book. Individual mapmakers of this period deserve more comprehensive study than the brief comments made here. A résumé of the career of one of them about whom little has been known may serve as an example.

CLAUDE JOSEPH SAUTHIER

A mapmaker of the revolutionary war period whose manuscript and printed maps are found not only in the Earl Percy Collection but in other libraries in Europe and America deserves special study for his contributions to North American colonial British cartography. He has been mentioned in the previous chapters, since his range of activity is varied in type and includes work both in the

South and in the North; his major work, however, was called forth by the Revolution.

Claude Joseph Sauthier was an expert draftsman and surveyor, one of the most productive cartographers of the American scene before 1800. Although some of his work such as his "Chorographical Map of the Province of New York" (1779) is well known, he himself has remained a shadowy figure about whom few details have come to light. He arrived in America in 1767 as surveyor and draftsman for Governor Tryon of North Carolina. In the fall of 1768, he began a series of surveys of ten towns in North Carolina. The plans of Hillsborough, Brunswick, Edentown, New Bern, Halifax, Bath, Wilmington, Beaufort, Cross Creek (Fayetteville), and Salisbury, set in their environs, with the location of public and private buildings, gardens, farms, and roads, are unsurpassed in any other colony or province in number, detail, and charm of execution.[75]

A reference to himself as a native of Strasbourg, France, made by Sauthier in a loyalist claim in the Audit Office files in the Public Record Office, has led to other facts about him.[76] Sauthier was born in Strasbourg on 10 November 1736, the son of a saddler, Joseph Philippe Sauthier, and his wife, Barbe Primat.[77] He was one of a large family, the most distinguished of whom was his younger brother, Joseph Philippe Sauthier, who was born in 11 July 1751. Joseph took doctorates at the episcopal University of Strasbourg in philology and in theology, and became professor of logic in the university at the age of twenty-four. Later he became professor of theology in the Grand Séminaire, was appointed to several other offices, and died in Paris as a canon of Saint Denis in 1830. Claude Joseph, our particular interest, studied architecture and surveying; the preservation of his drawings and maps in the Grand Séminaire in Strasbourg may be owing to his younger brother's long connection with that institution. Claude Joseph grew up during a period of cartographic activity in Alsace; the precise system of triangulation established in 1744 for mapping all France marked a new epoch in which many local surveyors were needed and employed. In 1759 de Luche, superintendent of Alsace, began a compre-

hensive and detailed survey of that part of France with cadastral mapping of agriculture, forest cover, pastures, and other features.[78] Whether Sauthier was employed in these surveys or not, their methods influenced his maps, as examples of his work in the library of the Grand Séminaire show. He also finished in 1763 a work on civil architecture and landscape gardening, with beautiful colored designs, including some for a governor's mansion.[79]

It may be that Governor Tryon knew of these productions when he employed Sauthier in 1767. At the time he was building a governor's mansion at New Bern that, when finished, was to be the finest in the American colonies. Sauthier's plan of New Bern, surveyed and drawn in 1769, shows elaborate formal gardens between the mansion and the Neuse River. The gardens may not have been completed by 1769;[80] is it possible that Sauthier inserted a design that he had been told to plan and execute? In May 1771, Sauthier accompanied Governor Tryon to Alamance, where provincial insurgents called the Regulators were defeated; Sauthier drew four maps of the battle.

Later in 1771 when Tryon was appointed governor of New York, he took his surveyor with him. Sauthier, during the next few years before the Revolution, was occupied in extensive surveys of the province and of the City of New York. His first map of the province, including surveys by Bernard Ratzer as well as his own, was published in 1776; his great "Chorographical Map of New York" appeared in 1779. In 1773 Tryon appointed him surveyor for the Province of New York to run the boundary line between New York and Quebec.[81] About this time he acquired five thousand acres from Governor Tryon in the township of Norbury, Vermont.[82]

These peaceful occupations and preoccupations of Sauthier came to a sudden end. With the coming of the Revolution he became a military draftsman and surveyor. In 1774 he had accompanied Tryon to England; he returned with him to New York in 1775 to turmoil and uprisings. After General Howe landed his troops from Halifax on Staten Island in 1776, he ordered Sauthier as a military surveyor to map Staten Island. At Alnwick Castle is that very map, which came eventually to Percy's hands; it is interesting to surmise that the heavy stains on it are the result of field use by Earl Percy or General Howe on Staten Island before they moved over to Long Island.

When Earl Percy made his attack on Fort Washington, Sauthier on the day of the battle surveyed the field and made a map, now in the Library of Congress.[83] Percy evidently liked Sauthier and his work; he took him along to Rhode Island, where Sauthier drew a map of Newport, where Percy had made his headquarters.[84] In 1777, when Percy returned to England, Sauthier again accompanied him and became his private secretary. Sauthier's days as an active military surveyor were over; but between 1785 and 1790 Percy, after he became the second Duke of Northumberland, had him make.several estate maps, some of which are preserved in the duke's London residence, Syon House.

Sauthier returned in the end to Strasbourg, where he lived at 14 de la Grand' rue until his death at age sixty-six on 26 November 1802. There in the library of the Grand Séminaire, where his brother was professor of theology, are Sauthier's work on architecture and a few of his printed and manuscript maps of American and European subjects.

CONCLUSION

The British, having defeated the French in the struggle for America and then lost it to the Americans, went home. Many an officer's trunk was crammed with the maps and charts of American lands and shores that he had used or made. These have found their way into the attics of great houses and into national archives. It is quite possible that others will be discovered.

What was left in America of the result of the labor and skill expended by the British in the mapping of this land in the eighteenth century? First, many copies of these remarkable maps remained to be used as a foundation for American cartography. Second, much skill was learned from the British by Americans who had themselves been British.[85] Third, a standard for mapmaking was established that is comparable to the standards for education and government left in other parts of the empire from which Britain has withdrawn. And, fourth, a far greater knowledge, although still imperfect, of this enormous land was accumulated than would have been possible without the achievements of the eighteenth-century British colonial cartography of North America.

Appendix A

American Manuscript Maps of Sir Francis Bernard

These maps form part of a large collection of eighteenth- and nineteenth-century documents of the Bernard family The owner and custodian is Dr. J. G. C. Spencer Bernard, Nether Winchendon House, near Aylesbury, Bucks. The Historical Manuscripts Commission listed the entire collection in 1958–59: H.M.C., *National Register of Archives*. "Report on the Spencer Bernard MSS. February 1960." *BS* 33/79 (1103). A typed copy of the report is available at the National Register of Archives, Quality Court, Chancery Lane, London W.C. 2. The identifying numbering affixed to the maps by the H.M.C. is given in the following list.

Some of the maps are not described here fully because the sheets originally joined have become unglued and detached or have insufficient identifying information; a few are so fragmented that attempts to measure them for size and scale were futile. The map collection is to undergo preservation work. Printed maps are not included in this list, since those in the collection are not rare.

Map of Sir Francis Bernard's American Estate & The Adjoining Country [cartouche, bottom right].
Size: 25″ x 28″.
Scale: 1″ = ca. 9 miles.
Description: This beautifully drawn map, in black ink with bays, rivers, and lakes colored in green, shows the coast from Broad Bay past Penobscot Bay and its islands, to "Passamaquiddy Bay" and Saint Croix River. The shoreline from Mount Desert Island to Passamaquoddy Bay is unmarked. The course of the Saint Croix River to the Grand Lakes, the Penobscot to Chesuncook Lake, and the upper reaches of the Kennebec River to Moosehead Lake are traced carefully, with the beginning of the overland trail to the River Saint Lawrence.

To His Excel^y Fran^s. Bernard Esq^s, Captain General & Governor in Chief in & over his Majestys province of ye Massachusetts Bay in N Engl^d and Vice Admiral of the same. This Plan Of y^e Town of Pownall by order of y^e propietors of y^e Kennebeck Purchase from y^e late Colony of new Plymouth . . . James Pitts James Boydoin Benj^m Hallowell Thos Hancock and Silv; Gardiner / Boston 12th Novr. 1763 / Tho Johnston Sculp [cartouche, center left].
Size: 30″ x 21½″.
Scale: 1″ = ca. 2570 feet.
Description: A colored map of the area bordering the coast from the Kennebec River to "Witcheassett Bay" and "Sheepscutt River." It is a detailed survey of land ownership, with lots stretching back from the Kennebec River and the bay, ranging from 40 acres to 2,200 acres, with drawings of houses already built, and with Richmond Fort on the west bank of the Kennebec River.

MP / 19

John Small. To his Excellency Francis Barnard Esqr. Capt Genl and Governour in Chief in and over his Majesty's Province of the Massachusetts Bay in New England Sir, Please favourably to accept this Map of the County of York Cumberland & Lincoln Taken from the best Au-

thorities by your Excellencey's Most Obed^t. and very humb^{le} Servt John Small Scarborough April 10th 1761.
Size: Four sheets, each 19¾″ x 15⅜″.
Scale: 1″ = 5 miles.
Description: The coast from Portsmouth (with neatly drawn figures of individual houses) to Penobskeag River, with inlets, rivers, and offshore islands in great detail. The rivers were evidently not surveyed more than thirty miles from the coast except for the Kennebec River, which extends to the top of the map, to the indication "from this pond is Suppos'd to be a Short Carrying place into Penobscot River." Along this coast are townships such as York, Kittery, Wells, and Falmouth, from the Piscataqua River to the Androscoggin River. There is a fort near the mouth of the Androscoggin. Straight lines northwest separate the different counties: York, Cumberland, etc.

MP / 11
A Copy of a Plan of the Six Townships Laid Out on ye East Side of Mount Desseart now called Union River. Taken partly in the Year 1763 and Finished in the Year 1764; as also a Plan of the Land Laid out Six Miles Latetude, above the North Line of the Aforesaid Six Townships; Together with a Small Township Lying on the North Side of N°. 3 and on the South of the Six Miles Latetude . . . Taken Novr. y^e 12th, 1764, for Joseph Frie.
Scale: 1″ = 1 mile.

MP / 25
[Kennebec River with the Sagadehock or Amorescoggin branch].
Size: 19″ x 27″.
Scale: not indicated.
Description: On the Kennebec River above Swan Island are shown: Fort Francford, Fort Weston, and Fort Halifax. By Fort Halifax is a legend, "Here begins Captain Small's survey" [See MP/19]. Along the upper reaches of the river to the mountains are symbols for settlers' houses; Indian crossings are also indicated. At the top right of the map is an empty cartouche.

MP / 10
[Coast of Maine, north of New Hampshire] surveyor, F. Miller.
Size: Two sheets, 29½″ x 21½″ and 29½″ x 19¾″.
Scale: 1″ = 5 miles.
Description: A detailed, finely colored map of the coastal area from Portsmouth to Penobskeag River, including Casco Bay and Penobskeag Bay, with the county boundaries marked.

MP / 12
[Plan of lower Penobscot River and adjacent land].
Size: 20½″ x 29¾″.
Scale: 1″ = 1 mile.
Description: Colored map with a section of the Penobscot River at the top, Fort Pownall to the left, and Neskeag Point at the bottom center. Symbols indicate types of forest cover.

MP / 26
[Coast from Elizabeth Cape to George's River].
Size: 19″ x 7¼″.
Scale: not indicated.
Description: A coastal map of rivers, inlets, and islands from Cape Elizabeth by Falmouth Neck and Stroud Water to George's River.

MP / 27
[Part of Maine Coast].
Size: 20½″ x 29″.
Scale: 1″ = 1 mile.
Description: From Gouldsborough Harbor on the left northeast to Little Manana, Pigeon Hill, and Shipstern Island, with lines for allotments of property Nos. IV, V, and VI.

MP / 17
Indian Draught of the River St Croix i.e. the most easterly river in Passimaquoddy Bay [top left].
Size: 16″ x 10″.
Scale: not indicated.
Description: The Saint Croix River, from its mouth in "Passimaquoddy Bay," with the falls and portages between lakes. At the top are a river and lakes with the note, "probably a branch from St Johns."

MP / 7
A Plan of the Road between Boston and Penobscott Bay Surveyed by order of the Governor in pursuance of a Resolution of the General Court of the Massachusetts Bay. by F. Miller [1765; cartouche on middle sheet].
Size: Three sheets, each 26¾″ x 18¼″.
Scale: 1″ = 2 miles.
Description: The road from Boston to Penobscot Bay, Saint George's River. This map is in excellent condition, with coloring; there are symbols for county towns and churches or meetinghouses. By the compass declination is a note, "Marks the Course of the Magnet in 1765."

MP / 4
[Sketch map of the road from Boston to Portsmouth, 67 miles, and from Boston to Weston, 15 miles].
Size: 26¾″ x 18¼″.
Scale: 1″ = ca. 2 miles.

MP / 9
[Sketch map of route from Boston eastward to North Yarmouth, numbered 1–150 miles].
Size: Ten sheets, each 25″ x 14½″.
Scale: 1″ = ca. 2 miles.

MP / 1
[Route map from Brunswick on Stephens's River past Kennebec River and past Saint George's River to Penobscot Bay].
Size: 70⅞″ x 14½″.
Scale: 1″ = ca. 1¼ miles.
Description: A survey of the coastal road from 150 miles north of Boston to 211 miles north of Boston.

MP / 2
[A sketch map of the road from near Brunswick along Penobscot Bay to Saint George's Fort on Saint George's River, 210 miles north of Boston].
Size: 25″ x 14½″.
Scale: 1″ = 2 miles.

MP / 3
[A sketch map of the Boston road from Kittery past Wells, Biddeford, and Falmouth to New Casco, from mile 68 to mile 137].
Size: 25″ x 14½″.
Scale: 1″ = ca. 2 miles.

MP / 21
[Portion of the road to Boston by way of Springfield, numbered miles 1–68].
Size: Two sheets, each 26¾″ x 18¼″.
Scale: 1″ = ca. ⅔ miles.

MP / 22
[Part of a route map west of Springfield through Great Barrington, Massachusetts].
Size: Six sheets, each 26¾″ x 18¼″.
Scale: 1″ = ca. ⅔ miles.
Description: Includes part of the Boston to Albany, New York, road num-

bered from mile 33 to mile 111. Along the route are posthouses such as Burchhardt's Tavern and Whiteing's Tavern; in Great Barrington the courthouse, church, and meetinghouse are located.

MP / 23
[Road map from Boston to Albany].
Size: Five sheets, each 26¾″ x 18¾″.
Scale: 1″ = ca. ⅔ miles.
Description: This section of the Boston to Albany route survey includes the Connecticut River and the Northampton courthouse, with houses, taverns, and intersections of crossroads. It is in excellent condition, colored, with drawings of neighboring mountain ranges such as the Hadley Mountains.

MP / 24
[Part of the road between Boston and Albany, by Francis Miller].
Size: Four sheets, each 26¾″ x 18¼″.
Scale: 1″ = ca. ⅔ miles.

MP / 13
[Survey plan of Saint George's River (Penobscot Bay), by John North. 1759].

MP / 14
[Plan of Twelve Townships East of Penobscot River. By John Jones of Dedham].

MP / 6
[Charts of Passamaquoddy Bay].

MP / 28
Bundle of unidentified maps.
Bookwebweek River. Deck West River. Squogus River.

MP / 18
Unidentified maps.

MP / 20
[Chart of Moose Island and Cobscook Bay].

MP / 5
[Plan of part of the lands belonging to Brigadier Waldo; Moose Island].
Description: Samuel Waldo (1695–23 May 1759), whose "career is significant mainly for his unwearied efforts to develop his wild lands on the

coast of Maine between the Muscongus and Penobscot Rivers" (*Dictionary of American Biography* 19:333), was second in command of Massachusetts forces in the campaign against Louisbourg in 1745, with the rank of brigadier-general. The Waldo Patent embraced a tract thirty miles square, to which he attracted by generous offers Scotch-Irish settlers to Saint George's (Thomaston), English to Medumcook (Friendship), and Germans to Broad Bay (Waldoborough). He died near present Bangor, Maine, while on an expedition to establish a fort on the Penobscot River (Cyrus Eaton, *Annals of the Town of Warren in Knox County, Maine, with the Early History of St. George's, Broad Bay, and the Neighboring Settlements on the Waldo Patent* [2d ed; Hallowell: Masters & Livermore, 1877]).

EM / 2
[Plan of lands belonging to the heirs of Brigadier Waldo. Surveyor, Joseph Chadwick. 1765].

Description: The Waldo Patent, of well over half a million acres between the Muscongus and Penobscot rivers in Maine, was inherited by his two sons and two daughters, with a double portion for the elder son. Governor Francis Bernard, whose own lands were nearby, was much interested in attracting settlers to the region.

MP / 8
Part of the Eastern Main of Hudson's Bay, otherwise called Terra Labrador, extending Northward from the Streights of Belle-Isle, partly inhabited by the Eskimaux Indians.

This Plan is laid down from the Explorations of Capt Henry Atkins of Boston New England made in 1758 [cartouche, top left].

Drawn from Capt Atkins Original Plan by J. Leach [small cartouche to right of main cartouche].
Size: 58⅝″ x 10⅝″.
Scale: 1″ = ca. 6⅔ miles.

Appendix B

North American Manuscript Maps of the
Revolutionary Period from the Collection of
Earl Percy in Alnwick Castle

The manuscript and printed maps belonging to Hugh Earl Percy (1742–1817), later second Duke of Northumberland, were not calendared with other Northumberland documents by the Historical Manuscripts Commission. The printed maps in the collection, not included in this list, are chiefly those of Jefferys, Faden, and Des Barres. Among them is a first state of the first plate of Fry and Jefferson's "Inhabited part of Virginia . . . 1751 [1753]," of which only two other copies and a fragment are known.

After Earl Percy's arrival in Boston on 4 July 1774 with the Fifth Regiment of Foot of which he was colonel, General Gage placed him in command of British troops in Boston with the rank of brigadier-general; on 10 July 1774 he was appointed major-general in the British army in America. He took an active part in the campaigns in New England and New York. He returned to England from Newport, Rhode Island, where he was in command, on 5 May 1777. He did not come back to America; but his continued interest is shown by manuscript and printed maps made for him or collected by him throughout the course of the war. While in America he was able to draw on the services of some of the best mapmakers working for the army such as John Montresor, Edward Barron, and C. J. Sauthier; he took Sauthier with him to England as his secretary when he left Newport.

CANADA

Arrangement of the Army Commanded by General Howe. Agreeable to the Orders of the 15th May, 1776 [cartouche, bottom center]. EB [lower right corner of cartouche].

Size: Oval map 12 15/16″ x 9¾″ in shaded frame: 14 15/16″ x 11 11/16″.
Sheet: 17⅞″ x 15″.
Scale: not indicated.
Description: This colored oval plan, set in a shaded grey frame, is by Edward Barron of The Fifth Regiment of Foot (King's-own Regiment), who made several maps for Earl Percy in 1775 in Boston and a plan of the Battle of Camden, South Carolina, in 1780. In the Muniment House of Syon House, Islington, is a manuscript "Plan of the City and Castle of Chester," signed EB and dated 1786. This, with the manuscript map of the battle of Camden, may indicate that Percy employed Barron after the war.

On 15 May 1776, General Howe was able to make undisturbed this elaborate "Arrangement" of his army, deployed in battle order, while he was in Halifax, Nova Scotia. He had evacuated Boston on 17 March, after Washington's fortification of Dorchester Heights and the foiling of Percy's counterattack by a "hurrycane." He sailed north in one hundred and twenty transports carrying about ten thousand troops and dependents and eleven hundred Loyalists with their belongings. In Halifax, safe but overcrowded, called by an officer "a cursed, cold, wintry place," Howe sorted refugees, amassed stores, received reinforcing regiments whose colors are added here, and exercised his troops "in line," preparing for the July battle for New York that culminated in the taking of Fort Washington.

Reproduction: W. P. Cumming and E. C. Cumming, "The Treasure of Alnwick Castle," *American Heritage* 20 (1969):32.

Line of March of the First Brigade, from the Right by Sub Divisions. E. Barron [cartouche, bottom right center].
Size: 17 11/16″ x 11⅜″.
Scale: not indicated.
Description: This colored plan is undated, with no indication of location; it may be a companion piece to the preceding map.

Map of Nova Scotia drawn from the Original in the Office of the Surveyer-General of the Province. And Humbly Presented to the Right Honorable Earl Percy By His Lordship's Most Devoted Humble Servant Edwᵈ Barron Captⁿ Kings-own Regiment [top left].
Size: 35″ x 22⅜″.
Scale: 1″ = ca. 9½ miles.
Description: This colored map in green and brown, with towns in red, may have been made when Earl Percy was in Halifax in 1776. It was more probably made at the same time that Barron made other maps (listed below) for Earl Percy when Barron was in Halifax in 1779, surveying under the commanding engineer William Spry (see Christian Brun, *Guide to the Manuscript Maps in the William L. Clements Library* [Ann Arbor: University of Michigan Press, 1959], nos. 20, 22–24, 30).

Plan of the Town & Environs of Halifax in Nova Scotia, Drawn on the spot, and Humbly Presented to the Right Honorable Earl Percy, by His Lordship's most obiged [*sic*] and Most Devoted Humble Servant Edw. Barron. 1779 / References [right side].
Size: 31 15/16″ x 21¼″.
Scale: not indicated.
Description: This plan shows the houses and defenses of the town. It is beautifully drawn in color in Barron's usual style, with numerous trees each casting its shadow.

Plan of Fort Cumberland On The Isthmus of Nova-Scotia, 1779 [cartouche, bottom left].
Size: 18 13/16″ x 12¼″.
Scale: 1″ = 60 feet.
Description: This colored plan of the fort is probably by E. Barron.

Plan of Fort Edward at Windsor Nova-Scotia, 1779. E Barron Fecit [cartouche, top right]. / References, A–H [bottom right].
Size: 17⅝″ x 10½″.
Scale: 1″ = 60 feet.
Description: This colored map of the fort has outlying buildings and hachuring of elevations.

Rough Sketch of Colean du Lac by a Scale of 40 feet to an inch.
Size: 17 7/16″ x 12½″.
Description: A colored plan of barracks and a canal. Another sheet has "A Plan and Section of a Blockhouse proposed to be built at Colean du Lac" with an accompanying letter from Mr. Glenie explaining and describing it.

MASSACHUSETTS

[Road map, untitled, from Boston Neck past Brookline around to Cambridge, and with adjacent roads].
Size: 62⅜″ x 27⅝″.
Scale: not indicated.
Description: This large pen-and-ink sketch, with roads crudely drawn, shows houses with names of owners along the main and adjacent routes. Two buildings of Harvard College are carefully drawn, with the jail across the street from the college.

[Crude pen-and-ink road map: Cambridge is at the top left, Menotomy meetinghouse at the top right; Charlestown is at the lower left and Medford meetinghouse at the bottom right, with intersecting roads].
Size: 23⅜″ x 15⅛″.
Scale: not indicated.
Description: This topological sketch, on two folio-sized sheets pinned together, has numerous legends describing the terrain: "hilly broken ground but no trees," "flat and open ground," "Each Side lined with Trees."
From a "Tavern" on the Menotomy-Cambridge "great road" a curving road leads to the "great road to Charlestown [from Cambridge]," with the accompanying note, "Kent's Lane through which the Troops returned from Concord." In all the voluminous documents recording the events of 19 April 1775, this is apparently the only reference to Kent's Lane as the road taken by Earl Percy onto which he wheeled left to avoid the concentration of militia in Cambridge that probably would have destroyed his troops if he had continued his planned return. This is the only detailed map of this countryside among Earl Percy's papers, although unless the Kent's Lane notation was added later by the same hand, he did not have it with him on 19 April.
Reproduction: Cumming, "Alnwick Castle," p. 28, and fig. 32 in this volume.

A Plan of the Town and Harbour of Boston and the Country adjacent with the Road from Boston to Concord. Shewing the Place of the late Engagement between the Kings Troops & the Provincials, together with

the several Encampments of both Armies in and about Boston. 19th April 1775 [cartouche, bottom left].
Size: 11⅞″ x 7¾″.
Scale: 1″ = 2 miles.
Description: This unsigned manuscript map is the first known British map to show the retreat from Concord on 19 April 1775, although the rough pen-and-ink sketch of the smaller area east of Menotomy showing the return by Kent's Lane may be earlier. As the map indicates, the heavy attack began not with the shots "at the rude bridge that arched the flood" in Concord but at the Mill Brook bridge near Meriam's Corner, where the British under Colonel Smith fired a parting volley, and the watching militia closed in. There are inaccuracies on the map: Percy did not advance beyond Lexington, and he returned through "Menatony" (Arlington) toward Cambridge. The provincial army assembled around Boston with amazing speed; by 21 April about nine thousand surrounded the campfires seen by the British from Beacon Hill. The placing of encampments on this map shows that it was probably made at least by 26 April; on the 22d many regiments were at Watertown, but were ordered to Cambridge on the 26th, where headquarters were established by General Ward.

This copy of the map, belonging to Earl Percy, is the only one known; but another must have been sent to England and used as the basis for a more elaborate map with later and additional information that I. De Costa published in London on 29 July. The titles are identical, except that De Costa omits "19th April 1775" and adds "Taken from an Actual Survey Humbly Inscribed to Richd. Whitworth Esqr. Member of Parliament for Stafford. By his most Obedient servant I: De Costa, London, July 29th, 1775," with a table of references below. De Costa was helped by Jonathan Carver, according to a letter written by Is. Foster to Major Robert Rogers on 8 August 1775: "Carver and DaCosta have finished a new plan of Boston at the request of Whitworth." Elsewhere in the letter Foster mentions that Carver expected to be appointed Superintendent of Indian Affairs through Whitworth's influence: see Historical Manuscripts Commission. *Dartmouth Mss.* 2 (14th Report, App., Part X):350.

Although the center of De Costa's map shows close dependence on a plan identical with Percy's, it must have drawn on some other source for its enlarged area, from "Bellerika" and "Salem" in the north to "Nantick" and "Weymouth" in the south, with the inclusion of the Boston outer harbor to the east. Percy's map employs colored rectangles for troops in military engineering usage; De Costa decorates his colored figures of soldiers to give commercially effective pictorial attractiveness. De Costa adds details concerning the Battle of Bunker Hill on the map, with identification of the naval ships in the harbor in the table below the title. This was hot news and must have required rapid work for the designer and for C. Hall, the engraver; General Gage's report of the battle had appeared in the *London Gazette* on 25 July, only four days before the publication of the map. Three different states have been noted. The earliest has "2. The Somers [? Somerset] Man of War" in the Table of References. A collotype reproduction of the Yale University Library copy of this state was made by the Meriden Gravure Company in 1963. A second state changes the Somers reference to "2, The Lively Man of War," and the location of "Charles Town." A reproduction is in E. D. Fite and A. Freeman, *A Book of Old Maps* (Cambridge, Mass.: Harvard University Press, 1926), pl. 64 (reproduction of John Carter Brown Library copy; Dover reprint, 1969, Library of Congress copy). A third state adds the name of General Washington at the provincial headquarters. Washington took over command of the army on 3 July 1775. An example of this state is in the Newberry Library.
Reproduction of Percy MS: title page this volume.

A Draught of the Towns of Boston and Charles Town and the Circumjacent Country, shewing the Works thrown up by His Majesty's Troops; and also those by the Rebels during the Campaign, 1775. N.B. the Rebels entrenchments are express'd as they appear from Beacon Hill which are colored yellow. [In lighter ink:] John Montresor commandg. engr. to Major General Earl Percy [cartouche, top left].
Size: 17⅝″ x 17⅛″.
Scale: 1″ = 2,190 feet.
Description: This finely designed and colored map of Boston and vicinity extends from "Mistick River" south to "Thomson's Id." It was made during the summer or fall of 1775. John Montresor (1736–1799) accompanied Earl Percy on 19 April and supervised the rebuilding of the bridge over the Charles River on the march to Lexington. On 18 December 1775, Captain Montresor was made chief engineer of British forces in America by George III.

A Plan of the Town of Boston, 1775 [cartouche, lower right].
Size: 26¾″ x 18⅝″.
Scale: 1″ = 432 feet.
Description: This black and white drawing, with faint blue coloring for edges, gives the names of streets and wharves, etc.

A Plan of Boston Advanced Lines. 1775 [bottom right].
Size: 13¼″ x 13⅞″.
Scale: 1″ = 400 feet.

Plan of the Peninsula of Charlestown. E. Barron [cartouche, lower left].
Size: 20¼″ x 15″.
Scale: 1″ = 400 feet.
Description: A colored plan of the two forts and the location of British troops on the peninsula after the battle of 17 June 1775. Although Earl Percy's troops took part in the fight, he was in charge of the defenses at Boston Neck that day.
Reproduction: Cumming, "Alnwick Castle," p. 30.

A Sketch of Charlestown Neck, &c., &c. [title, top right].
Size: 20 3/16″ x 13⅞″.
Scale: not indicated.
Description: This colored map in green and brown, with red dots for the "Ruins of Charles Town," was apparently made soon after 17 June 1775, since no fortifications are on Bunker Hill or Breed's Hill. Offshore are two rowboats and a ship. The country to the north of Charlestown Neck is shown, with the roads to Medford past Winter Hill and to Salem by Mount Pisgay.

A Sketch of the Peninsula of Charles-Town with Gen. Howe's Intrenchment and the Rebels Redoubt.
Size: 10 13/16″ x 8 15/16″.
Scale: not indicated.

View of Roxborough Church, Kings Arms, & the Chimnies of Brown's House: from an Embrasure between the Sally Port & the right hand Bastion of the Lines. S. Biggs 1775 [title legend below neat line].
Size: 14 1/16″ x 9⅞″.
Size of sheet: 15⅜″ x 12⅛″.
Scale: not indicated.
Description: A pen-and-ink drawing, uncolored.

Dorchester Point Copied from a Actual Survey [center; legend with no cartouche].
Size: 7¼″ x 9″.
Scale: not indicated.
Description: This is a brown ink sketch with Boston at the top and Castle William Island at the lower right. The legend, "Shows where the Rebels have a guard," accompanies dotted positions on Dorchester Point.
  This map was drawn before the American fortifications of Dorchester in 1776. Between 4–17 March 1776, George Washington built redoubts, not shown on this map, which forced Howe's withdrawal of British troops from Boston by making it untenable. Burgoyne had earlier urged the

British to fortify the Point, but in vain. Howe ordered Percy to attack Dorchester Heights; but very inclement weather prevented the proposed action, and on 17 March Howe, with all the troops and the Loyalists with their families and household goods, sailed for Halifax.

RHODE ISLAND

A plan of Rhode Island and the Adjacent Islands and Shores.
Surveyed in the Year 1774 Under the Direction of Samuel Holland [top left; no cartouche], George Sproule Lieut 16 Foot [bottom right].
Size: 50¾″ x 39⅜″.
Scale: 1″ = 2,000 feet.

Plan of the Town of New Port with its Environs. Survey'd by Order of His Excellency the Right Honourable Earl Percy Lieutenant General Commanding His Majesty's Forces on Rhode Island &c. &c. &c. In March 1777. By C, J, Sauthier [cartouche, bottom right, with references 1–12 below title above signature].
Size: 28⅝″ x 38⅜″.
Scale: 1″ = ca. 512 feet.
Description: This detailed map shows houses in red, lots in green, with roads and fields and hachured elevations. Earl Percy sailed for England on 5 May 1777, taking with him as his secretary C. J. Sauthier.

A Plan of the Adjacent Coast to the North Part of Rhode Island, to Express the Route of a Body of Troops, Detatched to Destroy the Rebels Batteaux, Vessels, Stores, &c., &c. Accomplished May 25th 1778. [In cartouche, written in a different hand: Drawn by M. Seix (?) Capt: in 22ᵈ Regˢ] [cartouche across entire bottom].
Size: 16 11/16″ x 14⅝″.
Scale: 1″ = 1 mile.
Description: This colored map shows Mount Hope Bay and Taunton River to the east, Prudence Island to the west, and the entrance of Providence River to the northwest.

A Plan of the Adjacent Coast to the North Part of Rhode Island to Express the Route of a Body of Troops, detached to destroy the Rebel's Batteaux, Vessels, Stores, &c. &c. Accomplished May 25th 1778. Under the Orders of Lt. Col: J. Campbell 22ᵈ Regt [cartouche, top right].
Size: 13½″ x 16 13/16″.
Scale: 1″ = 4,224 feet.
Description: This colored map, with hachures of elevations, gives details of troop and naval movements in red. Bristol Neck is in the center of the map. In Mount Hope Bay are drawn "Six boats proceding from [Pigot]

the Royal Galley to take the Enemy" and at the mouth of Taunton River "Rebels Galley Taken from the mouth of Falls River." Numerous legends make this an interesting map. The destruction of their sawmill on Falls River made the Battle of Falls River a serious blow to the Americans; it set back their plans for building batteaux to invade Rhode Island.

Plan of the Southern Part of Rhode Island with the Harbour, and such of the Adjacent Coasts as is necessary to express the entrance of the Three Rivers into the Narragansett Bay [cartouche, top center]. The Plan of the Southern Part of Rhode Island is laid down from a very late survey, to exhibit the ground round the Town of New-Port, and the Situation of the works and Line of Defense occupied by the British Troops when attacked in August 1778; Also the Enemys Line of Approach by Land, with the Disposition and movement of the French Fleet by Sea [cartouche, top center] / Laid down from a Survey by Edwd. Fage, Lieu^t Royal Artillery Jany. 1779 [to left of cartouche].
Size: 33 3/16" x 25 3/16".
Scale: 1" = 2,000 feet.
Description: Lieutenant Edward Fage's map shows the town of Newport and southern Rhode Island with houses and streets in great detail; military and naval locations and accompanying legends are in red with water in blue and beaches in yellow.

NEW YORK
A Topographical Sketch of New York & Y^e: adjacent country [cartouche, top left].
Size: 43" x 22¾".
Scale: 1" = ca. 2,112 feet.
Description: This colored map of New York from the Battery to Dykemans gives "Road to Boston" with names of house owners above New York past Harlem. Two large sheets are pasted together for this sketch, which is carefully but not elaborately finished. It may have been made for Earl Percy at Halifax in 1776 in preparation for the attack on New York, by Captain Montresor, who made a survey of New York in 1766.

Map of Staten Island in the Province of New-York. Survey'd by Order of His Excellency General Howe, Commander in Chief of His Majesty's Forces in North America. August 1776. By C. J. Sauthier Ingineer Geographer [cartouche, top left].
Size: 35 1/16" x 27 13/16".
Scale: 1" = 2,112 feet.
Description: This careful survey of Staten Island gives roads, buildings, fields, and other details in color. It includes the entire island; adjacent parts of Long Island and New Jersey are included without detail.

The map is water-stained and mildewed; it probably saw field use during preparations for the campaign for New York. General Howe arrived off Staten Island at the end of June 1776 with his troops and the largest flotilla of naval vessels ever to have gathered in American waters. He landed on Staten Island and remained there until 22 August, when he began his campaign by sending the troops across to Long Island in a successful attack against the Americans that forced their retreat and evacuation.

Plan of the City of New-York as it was when his Majesty's Forces took Possession of it in 1776. Showing all the works the Rebels dit [*sic*] in the course of the preceding winter mark'd yellow and the part of the City which was burnt the same year by a red colour and dott'd lines. Survey'd in October 1776 by C, J, Sauthier.
Reference: 1–41 [e.g., 40, Niger Burying ground; 41, Jews Ditto].
Size: 31½" x 23½".
Scale: 1" = ca. 448 feet.
Description: This colored plan by Sauthier of the southern part of Manhattan Island, up to approximately the present 20th Street, is similar in area to Bernard Ratzer's "New York" (1767) but shows the extensive changes made by fires and fortifications. It includes part of Brooklyn on Long Island, with houses, orchards, and hachuring of elevations.

Among the Percy maps are four manuscript sheets relating to military activities on Lake Champlain from 1776 to 1778:
1 and 2. A Return of the Losses sustained by his Majesties Fleet on Lake Champlain, under the command of his Excellency Genl Sir Guy Carleton from the 30th of August 1776 to the 24th June 1777 [two large sheets describing losses of vessels and stores, with the cause and place of the losses and with a list of officers' names].
3. Men and officers in Dockyard at St. John's [with a list of vessels on Lake Champlain in 1778].
4. A Return of Officers now serving in Lake Champlain.

MIDDLE ATLANTIC COLONIES
[Delaware River from German-Town to Chester] N.B. The first attack was on Wednesday October 22nd 1777. And the Augusta, and Merlin were burnt on the 23rd d.[itto] the Second was on Saterday [*sic*] the 15th November D^to. [under reference list A to O, to right of map].
Size: 10⅞" x 14½".
Scale: 1" = 1 mile.

[Middle Atlantic Colonies- Army Campaign Movements, August 1776 through December 1779].

[**No. 1**] Reference of the Several Drawings Showing the Operations of the British Army Against the Rebels in North America from the 12th of August 1776 to the end of the Year 1779 [top of fol. *1a* of a large folded folio sheet: underneath, on *1a* are references numbered A to R, followed beneath by:-] N.B. The remainder of this year's campaign is on the Small Sheet with its Explanations. [Underneath, and continuing from *1a* through *1b*, *2a*, and *2b* are numbered items 1 to 63, with the following:] N.B. The Batallion[s] in red are the Kings Troops those in Green likewise only Shewing the Second Position, those in Yellow are the Rebels. The Side on which the Color Lyes Showes which way they face. [The preceding references, found on the large folded sheet, accompany three numbered sheets of different sizes, numbered but without titles, which are described below. The three sheets are numbered 2, 3, and 4; they are separate but, if put together, form a single connected map of the area shown. Since the reference lists mention only numbers 2, 3, and 4, No. 1 is the large folded sheet with the references.]

No. 2 [top center, above neat line].
Size: 27 1/16″ x 18 ⅜″.
Scale: 1″ = 2 miles.
Description: Represents lower New York, Staten Island, and northern New Jersey to the Delaware River. Trenton is included, with several ferry points on the river.

No. 3 [top center, above neat line].
Size: 41⅞″ x 22⅞″.
Scale: 1″ = 2 miles.
Description: Philadelphia is in the center of the map, which extends twenty miles north of Philadelphia, south to Wilmington, and forty-five miles east to thirty miles west of Philadelphia.

No. 4 [bottom center].
Size: 19⅞″ x 14 15/16″.
Scale: 1″ = 2 miles.
Description: The area represented extends from the mouth of the Sassafras River three miles below Salisbury, Delaware, north to Newcastle; it includes parts of Chesapeake Bay and Delaware Bay, showing Charlestown on North East River.

THE SOUTH

Sketch of the Disposition and Commencem^t. of the Action near Camden in South-Carolina 16th August 1780. As discribed in the Letters of the Right Honble Earl Cornwallis to the Secretary of State: and the Rebel Gates to Congress. Most respectfully inscribed to the Right Honble Earl Percy, by his Lordships most Humble Servant Ed. Barron [cartouche, top center] / E. Barron [bottom right].
Size: 1. Oval, 9 3/16″ x 6 15/16″.
      2. Design, 10″ x 7 13/16″.
      3. Sheet, 13 5/16″ x 9 9/16″.
Scale: not indicated.
Description: This colored plan of the order of battle, with smoke from cannon fire and delicate drafting in Barron's usual manner, is similar in form to his earlier plan of General Howe's arrangement of troops in Halifax (see the first item in the Alnwick Castle list, appendix B). The engagement near Camden resulted in the defeat and rout of the American forces under General Gates. At the top under the cartouche are the American forces: First Maryland brigade, main line forces, General Gist's brigade; at center, North Carolina militia; right, Virginia militia; extreme right, Colonel Armand's Legion, Continentals, Pettenfield's Corps. The British are below: right wing commanded by Lieutenant-Colonel Webster, left wing by Colonel Rawdon. Colonel Rawdon, later Lord Hastings of India, was an officer in Boston under Earl Percy, who showed in his letters a concern for young Rawdon's welfare and safety.

A manuscript map of the Battle of Camden among the Faden maps (no. 51) in the Library of Congress is reproduced in Winsor's *America*, 5:531. The Faden map, dated I March 1787, shows a larger area than the Barron plan, which is limited to the terrain showing the positions of the opposing forces. Three maps of the engagement are among the Clinton Papers (C. Brun, *Guide to the Manuscript Maps in the William L. Clements Library*, nos. 624, 625, 631).

From Barron's cartouche title, he drew his sketch from secondary sources at some later date. Cornwallis's reports of 20 August and 21 August were published in the *London Gazette Extra* for 9 October 1780.

Notes

1. Mapping the Southern British Colonies

1. "A discription of the land of virginia" (endorsed): London, Public Record Office, Maps and Plans G. 584; reproduction, W. P. Cumming, *The Southeast in Early Maps* (Chapel Hill: University of North Carolina Press, 1962), pl. 10, p. 118. White's manuscript drawings are in the British Museum, P & D 1906–5–9–1. L.B. 1 (1) and (2): reproduction, Cumming, *Southeast,* pls. 11 and 12.

2. Thomas Harriot, *A briefe and true report of the new found land of Virginia,* pl. 1. Harriot's work is Part I of T. de Bry, *Collectiones Peregrinationum in Indiam Occidentalem* (Frankfort, 1590–1634).

3. "A mapp of Virginia discouered to ye Falls, and in it's Latt: From 35. deg: & ½ neer Florida to. 41. deg: bounds of new England. [Top right] John Farrer Esq., Collegit. Are sold by I. Stephenson at ye Sunn below Ludgate. 1651 [bottom center] John Goddard Sculp: [bottom left]." For a reproduction of the fourth state and a discussion of the changes in the plate see Cumming, *Southeast,* pl. 29, p. 141–42. Lawrence C. Wroth provides an excellent account of the background of the map in John Carter Brown Library, *Annual Report* (Providence, 1949–50), pp. 15–26.

4. John Lederer, *The Discoveries of John Lederer.* Ed. by William P. Cumming (Charlottesville, 1958). Locke's map is Public Record Office M. P. 1./11.

5. "A New Map of the Country of Carolina. With it's Rivers, Harbors, Plantations, and other accomodations. don from the latest Surveighs and best Informations, by order of the Lords Proprietors. Sold by Ioel Gascoyne at the Signe of the Plat nere Wapping old Stayres. And by Robert Greene at the Rose and Crowne in ye middle of Budge Row [two lines crudely erased from the plate, probably referring to another bookseller]." The map was evidently sold as a separate sheet and with the promotional pamphlet, *A True Description of Carolina.* London: Printed for Joel Gascoin and Robert Greene . . . [1682]. See Cumming, *Southeast,* pp. 34–36, 159–60.

6. "This New Map of the Cheif Rivers, Bayes, Creeks, Harbours, and Settlements, in South Carolina Actually Surveyed . . . by Iohn Thornton. & Robt. Morden." See Cumming, *Southeast,* pl. 42 and pp. 36, 166–67. Mathews' large manuscript wall chart is British Museum Add. MS 5415. 24.

7. Lawrence C. Wroth's account of the Crisp map is in John Carter Brown Library, *Annual Report* (Providence, 1950–51), pp. 10–17; see also Cumming, *Southeast,* pp. 40–42, 179–80.

8. John Love, *Geodaesia: The Whole Art of Surveying and measuring of land made easie . . . Moreover, a more easie and sure way of Surveying by the Chain, than has hitherto been taught . . . as also how to lay out new Lands in America . . .* , 2d ed., with additions (London: W. Taylor, 1715), appendix, p. 7.

9. *Documents Relative to the Colonial History of the State of New-York.* Ed. by E. B. O'Callaghan and others (Albany: Weed and Parsons, 1853–61), vol. 5 (1855), p. 577.

10. Woodbury Lowery, *The Lowery Collection,* ed. by P. L. Phillips (Washington, D.C.: Government Printing Office, 1912), p. 230.

11. "Carte de La Mer du Sud & de La Mer du Nord . . . . Par N. de Fer . . . . Paris. 1713." This is a large map, 75″ x 47″; beautifully colored copies are in the Library of Congress and the Harvard Map Room. A reduced copy of the map is in Henri A. Chatelain, *Atlas Historique*, 7 vols. (Amsterdam, 1705–20), vol. 6 (1719), no. 30, p. 117.

12. R. P. Louis Hennepin, *Nouvelle decouverte d'un tres grand Pays Situe dans l'Amerique* (Utrecht: Guillaume Broedelet, 1697). The same work was published in English, with some additions, as *A New Discovery of a Vast Country in America* (London, 1698), with the plates re-engraved.

13. "A New Map of the North Parts of America claimed by France . . . . the South West part of Louisiana is done after a French Map Published at Paris in 1718 . . . 1720." In: H. Moll, *The World Described* (London, 1709–20), no. 9.

14. London, Public Record Office, Colonial Office 700, Nova Scotia 4. The title of this colored manuscript chart is in a baroque cartouche at the top left: "To the Rt. Honble the Lords Comrs for Trade and Plantacons &c This Plaine Chart of the Coasts on the Province of Nova Scotia et Accadia &c is most humbly dedicated & presented by Nathaniel Blackmore anno 1714/15." Professor Arthur H. Robinson, whose comments have been helpful, is making a fuller study of Blackmore's chart and the lines on it.

15. H. Moll, *Atlas Minor* (London, 1729), no. 54.

16. Don W. Thomson, *Men and Meridians*, 3 vols. (Ottawa: Queen's Printer, 1966), 1:89, 316 n. 6. (chap. 10).

John Mitchell's comments are in a note on the third state of his map: "The map of New England and Nova Scotia requires a farther Consideration as we find them erroneously laid down in all our Maps, copied from a New Map of Nova Scotia . . . those errors are maintained by Arguments and pretended authorities, which seem to have confirmed them. The only Authority they have for all this is a feigned Survey by a pretended Surveyor General Blackmore in 1711–12 who appears by his Journals to have been Lieut. of the Dragon Man of War 1711 and made a rude Draught of this Coast (as well as he remembered it perhaps) in 1715, with a Petition to the Board of Trade to enable him to Survey it at that time, which he never did as we can learn . . . Mr. Moll published it as an Actual Survey made by her Majesties especial Command from which this Coast has thus erroneously laid down ever since . . ."

Mitchell is apparently referring to "The New Map of Nova Scotia and Cape Britain . . . Composed from a great number of actual Surveys . . . 1755," by John Green (Braddock Mead), Thomas Jefferys's geographer. Green, in his *Explanation for the New Map of Nova Scotia and Cape Britain, with the Adjacent Parts of New England and Canada* (London: T. Jefferys, 1755), p. 4, lists the numerous sources he used in the con-

struction of the map, among them the surveys of Surveyor General Blackmore. He adds on p. 12 that Blackmore made the peninsula too narrow, Henry Popple too wide.

17. Edward McCrady, *The History of South Carolina under the Proprietary Government, 1670–1719* (New York: Macmillan, 1897), pp. 236, 456, 530. *South Carolina Chancery Court Records 1671–1779,* ed. by Anne King Gregorie (Washington: American Historical Association, 1950), p. 260. W. Roy Smith *South Carolina as a Royal Province 1719–1776* (New York: Macmillan, 1903), p. 162. Mrs. Granville T. Prior, director of the South Carolina Historical Society, called the writer's attention to John Beresford's office as surveyor general in 1695, 1697.

18. Sir Robert Montgomery, *A Discourse Concerning the design'd Establishment of a New Colony to the South of Carolina, in the Most delightful Country of the Universe* (London, 1717), opposite p. 10 (cf. Cumming, *Southeast*, p. 185).

19. "A View of Savanah [*sic*] as it stood the 29th of March, 1734 . . . Peter Gordon" (cf. Cumming, *Southeast*, p. 202). See also John W. Reps, *Town Planning in Frontier America* (Princeton: Princeton University Press, 1969), pp. 335–47.

20. The Barnwell map (Public Record Office. Colonial Office. North American Colonies. General. 7) is without title, author, or date; reasons for its ascription to Barnwell are given in Cumming, *Southeast*, pp. 45–47, pl. 48. The Barnwell map is the basis of the following later maps (or of the southeastern region of more general maps): Popple "British Empire" 1727 (British Museum Add. MS 23615, fol. 72; cf. Cumming, *Southeast*, p. 82, n. 172); Popple "British Empire" 1733 (cf. Cumming, *Southeast,* pp.198–200); M. Catesby "A Map of Carolina . . ." 1731 (cf. Cumming, *Southeast*, p. 197); Bull (*Southeast*) 1738 (MS., P.R.O., C. O. Florida 2; cf. Cumming, *Southeast*, p. 207); Bull-Verelst (*Southeast*) 1739 (MS in the John Carter Brown Library; cf. Cumming, *Southeast*, p. 209, and John Carter Brown Library. *Annual Report. 1945–46*, pp. 20–21); Barnevelt (*Southeast*) ca. 1744 (MS in De Renne Collection. Georgia University Library; cf. Cumming, *Southeast*, pp. 212–13); Mitchell, "A Map of the British and French Dominions in North America," 1755. Most of these maps add later political changes and new settlements to the details of the original Barnwell map, omitting many of its legends. The Mitchell 1755 map makes the best and fullest use of Barnwell; the best of the manuscript maps is Barnevelt, ca. 1744, although it is a careless and inaccurate copy, as the name "Barnevelt" for "Barnwell" illustrates.

21. "A Map of the British Empire in America with the French and Spanish Settlements adjacent thereto. by Henry Popple. C. Lempriere inv. & del. B. Baron sculp . . . . London Engrav'd by Willm. Henry Toms 1733." A reproduction, with introductory notes by William P. Cumming

and Helen Wallis, was published by Harry Margary, Lympne Castle, Kent, in 1972.

22. Baron Lahontan, *New Voyages to North-America*, 2 vols. (London: H. Bonwicke, 1703), 2:13–14. In vol. 1, p. 124, Lahontan delineates some Indian territory "by way of a Map upon a Deer's Skin; as you see it drawn in this Map."

23. Thomas Pownall, *A Topographical Description . . . contained in the [annexed] Map* (London: J. Almon, 1776), ed. by Lois Mulkearn (Pittsburgh: University of Pittsburgh Press, 1949), p. 126, reproduced from a copy with manuscript additions in Pownall's own hand.

24. British Museum. Map Room. Royal United Services Institution Collection. Indian birchbark map.

25. "A Map of South Carolina and a Part of Georgia . . . London . . . T. Jefferys . . . 1757"; see Cumming, *Southeast*, pp. 54–55, 227–28. An excellent recent assessment of De Brahm's contributions is in Louis De Vorsey, Jr., "William Gerard De Brahm: Eccentric Genius of Southeastern Geography," *Southeastern Geographer* 10 (1970): 21–29. Louis De Vorsey's recent *De Brahm's Report of the General Survey in the Southern District of North America* (Columbia: University of South Carolina Press, 1971) gives the first adequate biography of De Brahm and edits the *Report* with scholarly notes.

26. C. J. Plowden Weston, ed., *Documents Connected with the History of South Carolina* (London, 1856). De Brahm's two-volume "Report" is now in the British Museum, King's Manuscripts 210–11; he made a second copy, now in the Harvard College Library, to which he appended "A Continuation of the Atlantic Pilot" with a detailed account of his observations on the Gulf Stream and the declination of the compass made on two voyages across the Atlantic Ocean in 1771 and 1775, with a map. Ralph H. Brown, "The De Brahm Charts of the Atlantic Ocean, 1772–1776," *The Geographical Review* 28 (1938):124–32, analyzes De Brahm's contribution to the study of the Gulf Stream.

27. Cumming, *Southeast*, p. 56.

28. *Colonial Records of North Carolina*, ed. by William L. Saunders; 10 vols. (Raleigh: State of North Carolina, 1886–90) 7:861.

29. "Virginia. Descouered and Discribed by Captayn John Smith. Grauen by William Hole." In: John Smith, *A Map of Virginia* (Oxford, 1612).

30. The John Carter Brown Library copy of Herrman's "Virginia and Maryland" has a printed slip with the imprint of John Seller, lacking on the only other known copy in the British Museum. A facsimile of its copy, published by the John Carter Brown Library in 1948, is accompanied by a bibliography of the map. John Carter Brown Library has included in its reproduction of the Blathwayt Atlas (Providence, 1970),

pls. 16–17, a manuscript version of Herrman's map. Another similar manuscript is in the Public Record Office. Miss Jeannette Black of the John Carter Brown Library writes that these two manuscripts (JCBL and PRO) were apparently copied by the same hand about 1677 from a version of the map earlier than the printing of 1673.

31. "A Map of the Inhabited part of Virginia containing the whole Province of Maryland with Part of Pennsylvania, New Jersey and North Carolina. Drawn by Joshua Fry and Peter Jefferson in 1751 . . . Tho. Jefferys . . . London." For the North Carolina contributions of William Churton see Cumming, *Southeast*, pp. 52–56, 219–21. The best study of the different editions of the map is by Coolie Verner, "The Fry and Jefferson Map," *Imago Mundi* 21 (1967):70–94. Although dated 1751, the map was not published until 1753. Extant copies of the first state are rare; in addition to those in the New York Public and University of Virginia libraries is one in the Percy Collection, Alnwick Castle, Northumberland, found by the writer in 1968.

32. "A New and Accurate Map of Virginia Wherein most of the Counties are laid down from Actual Surveys . . . By John Henry . . . Engraved by Thomas Jefferys . . . London, February 1770." See Thomas Pownall, *A Topographical Description of the Dominions of the United States of America*, ed. Lois Mulkearn (Pittsburgh: University of Pittsburgh Press, 1949), p. 6. See Fairfax Harrison, *Landmarks of Old Prince William*, 2 vols. (Richmond: Dietz Press, 1924) 2:634–36; John Carter Brown Library, *Annual Report, 1940–41* (Providence, 1941), pp. 44–47. Although Thomas Jefferys entered the map February 1770 (1771 New Style), it may not have been published until later. Henry was writing letters on 8 April and 10 April 1771 to Thomas Adams in London about the size of the printing order and distribution: "Notes and Queries," *Virginia Magazine of History and Biography* 58 (1950):132–33.

33. Edward B. Mathews, *The Maps and Map-Makers of Maryland* (Baltimore: Johns Hopkins Press, 1898); and Edward B. Mathews, ed., *Report on the Resurvey of the Maryland-Pennsylvania Boundary Part of the Mason and Dixon Line,* vol. 7 (Baltimore: Maryland Geological Survey, 1908). Both volumes have numerous reproductions of colonial maps.

34. John Richard Alden, *John Stuart and the Southern Colonial Frontier* (Ann Arbor: University of Michigan Press, 1944), and Louis De Vorsey, Jr., *The Indian Boundary in the Southern Colonies, 1763–1775* (Chapel Hill: University of North Carolina Press, 1966).

35. On the identity and achievements of the South Carolina surveyor James Cook see chap. **3**.

36. De Brahm's detailed and very large (20 by 5 feet) manuscript map of East Florida is in the Public Record Office. Colonial Office. 700. Florida 3. His survey, the result of nearly six years' work by De Brahm

and his assistants, was apparently used by Bernard Romans in his chart of the coast of East and West Florida, published in Romans's *A Concise Natural History of East and West Florida* (New York, 1775). P. L. Phillips reproduced this rare map in his *Notes on the Life and Work of Bernard Romans* (Deland: Florida State Historical Society, 1924), map 1. For useful comments on Romans's map see W. Lowery, *Lowery Collection,* ed. by P. L. Phillips (Washington: Government Printing Office, 1912), pp. 370–71; and John Carter Brown Library, *Annual Report, 1940–41* (Providence: John Carter Brown Library, 1941), pp. 49–53. Elias Durnford's two-volume collection of 116 manuscript maps of surveys of West Florida and the present Alabama-Mississippi region, made between 1765 and 1768, is in the Public Record Office, Colonial Office, 5.603, 604. Other Durnford maps of this area are in P.R.O., C.O. Florida 40, 45, 49.

37. Other copies of the Stuart-Purcell maps are in the Public Record Office, British Museum, and William L. Clements Library. The general maps of the Southeast and detailed surveys made by John Stuart or under his direction began with his "Map of the Cherokee Country 1761" (British Museum Add. MSS. 14036 fol. e.) and his "Map of the Southern Indian District 1764" (British Museum Add. MSS. 14036 fol. d). An examination of the background, development, and significance of the Stuart-Purcell maps, with a reproduction of the copy in the Ayer Collection, Newberry Library, is in Louis De Vorsey, "The Colonial Southeast on 'An Accurate General Map'," *The Southeastern Geographer* 6 (1966):20–32. An excellent and useful list of manuscript maps and surveys of the Southeast in the third quarter of the century, especially those relating to boundaries and Indian territory, is given in Louis De Vorsey, *The Indian Boundary in the Southern Colonies, 1763–1775* (Chapel Hill: University of North Carolina Press, 1966), pp. 256–59.

38. *Before the Indian Claims Commission. Docket No. 73. The Seminole Indians of the State of Florida, Petitioner, v. United States of America, Defendant. Defendant's Requested Findings of Fact, Objections to Petitioners' Proposed Findings, and Brief.* Ramsey Clark, Assistant Attorney General (Washington, D.C.: Government Printing Office, 1963), pp. 67–73.

39. Bernard Romans, *A Concise Natural History of East and West Florida* (New York, 1775). Facsimile edition (Gainesville: University of Florida Press, 1962), p. 270.

40. Historical Manuscripts Commission, *Dartmouth MSS,* vol. 2 (Fourteenth Report. London, 1955), p. 178. For this reference I am indebted to Professor Louis De Vorsey.

41. Public Record Office, Colonial Office, North America General, 15, "A New Map of the Southern District of North America from Surveys taken by the Compiler and Others, from Accounts of Travellers and from Best Authorities, etc. etc. Compiled in 1781 for Lieut Colonel Thomas Brown, His Majesty's Superintendent of Indian Affairs etc . . . . Joseph Purcell, 1781."

42. George C. Rogers, Jr., *Evolution of a Federalist: William Loughton Smith of Charleston (1758–1812)* (Columbia: University of South Carolina Press, 1962), p. 100.

43. Jedediah Morse, *The American Geography* (London, 1789) (no. 14, "1788"), 1794 (no. 14, "1792"), and reference in Preface to Purcell.

44. Public Record Office, M.P.D. 2.

45. Register of Saint Philip's Parish, Charlestown or Charleston, 1754–1810. Compiled by D. E. Huger Smith and A. S. Salley, Charleston: South Carolina Society of the Colonial Dames of America, 1927. Mrs. Granville T. Prior, director of the South Carolina Historical Society, has kindly furnished me with this and other references to Purcell in Charleston records. De Brahm states in his *General Survey* that Joseph Purcell was trained as "Draughtsman, Mathematician, Navigator": Louis De Vorsey, ed., *De Brahm's Report* (Columbia: University of South Carolina Press, 1971), p. 185. See also: Henry A. M. Smith, "The Baronies of South Carolina," *South Carolina Historical Magazine* 15 (1914):5. References to Purcell's surveys and plats of "baronies" around Charleston are scattered through many volumes of the *S.C.H.M.*; he was still active as a surveyor in 1803.

2. MAPPING THE NORTHERN BRITISH COLONIES

1. "A Map of New Jersey by John Seller, Ja. Clerk Sculp:" (London, 1664), in Seller's *Atlas Maritimus* (London, 1675). The fourth state with the "Restitutio" view of New York has a new title, "A Mapp of New Jersey, in America, by John Seller and William Fisher" (London, 1677). A detailed examination of the changes is given in Tony Campbell, "New Light on the Jansson-Visscher Maps of New England," *Map Collectors' Series No. 24* (London: Map Collectors' Circle, 1965). Coolie Verner has found a hitherto unrecorded state of the map, with one panel and a different version of the printed text beneath, to fit the single-panel map in the Pepys Library of Magdalene College, Cambridge.

2. A reproduction of the "fourth" state of Seller's map, with an accompanying pamphlet, is published by the John Carter Brown Library (Providence, 1958).

3. See Nicolaus Joannis Visscher, "Novi Belgii Novaeque Angliae," with an inset, "Nieuw Amsterdam op t Eylant Manhattans," in *Atlas Contractus* (Amsterdam [1666?], no. 60) and other Visscher atlases. See I. N. P. Stokes, *The Iconography of Manhattan Island* (New York: Robert H. Dodd, 1915–28), 1, pls. 7-A, 8. See also Hugo Allardt, "Totius Neo-

belgii Nova et Accuratissima Tabula." The different plates and states of these maps, with reproductions, are given in Campbell, *New Light on the Jansson-Visscher Maps*. For the different plates and issues of the Seller map see also Henry Stevens and Roland Tree, "Comparative Cartography," *The Map Collectors' Series No. 39* (London: Map Collectors' Circle, 1967), p. 329.

4. Available to Mitchell and Evans were the results of boundary surveys made in the current litigations of the province: see *A Bill in the Chancery of New-Jersey* (New York: J. Parker, 1747), no. 2 [New Jersey]. James Alexander, surveyor general of New Jersey, who died in 1756, has made numerous boundary line surveys and collected astronomical observations of latitude and longitude; a "Projection of that Province" made by Alexander came into the hands of Thomas Pownall and was used by him in his 1776 revised edition of Evans's map: T. Pownall, *A Topographical Description of the Dominions of the United States of America*, ed. Lois Mulkearn (Pittsburgh: University of Pittsburgh Press, 1949), p. 7.

5. "Nor have I at any Time published or given my Consent to the publishing of any Plan, Map, or Survey now extant, that bears my Name. Sam. Holland"; cf. Pownall, (ed. Mulkearn), p. 8, n. 3.

6. William Faden, *The North American Atlas* (London: W. Faden, 1777), no. 22. Faden revised the plate in 1778, with changes in town locations and road routes "from several Military surveys generously communicated by Officers of the British Troops and of the Regiments of Hesse and Anspach." This map was reproduced in facsimile by the New Jersey Historical Society for the state's tricentennial in 1964.

7. Hazel Shields Garrison, "Cartography of Pennsylvania before 1800," *Pennsylvania Magazine of History and Biography* 59 (1935):255–83; this is a brief but able treatment of the subject. See also Homer Rosenberger, "Early Maps of Pennsylvania," *Pennsylvania History* 11 (1944):103–17.

8. Römer's maps are usually colored and artistically decorated with trees, animals, and figures: British Museum King's Topographical Collection CXXII.27 (Hudson River); King's Topographical Collection: CXXI.10 (New York to Ontario); King's Topographical Collection: CXX.39. a. and b. (Castle Island in Boston Bay); Public Record Office, C.O. 700 New York 13 (very similar to King's Topographical Collection CXXI.10); Public Record Office. C.O. 700. New York, 13-B (New York Harbor). For Römer's activities on Castle Island see Justin Winsor, *Memorial History of Boston*, 4 vols. (Boston: J. R. Osgood and Co., 1880–81) 2:101.

9. Stokes, *Iconography*, 2:340. "To His Excellency Sr. Henry Moore . . . This Plan of the City of New York . . . Bernd. Ratzer . . . Thos. Kitchin sculpt . . . Surveyed in the Years 1766 & 1767 . . . London. Jany. 12 1776 by Jefferys & Faden."

10. This map is found inserted in some copies of Thomas Jefferys, *The American Atlas* (London: R. Sayer and J. Bennett, 1776). Other maps of New York by C. J. Sauthier and "A Map of the Province of New York, with Part of Pensilvania, and New England, from an actual Survey by Captain Montresor . . . London, A. Drury . . . 1775" are in William Faden, *The North American Atlas* (London, 1777).

11. Lewis Evans, *Geographical, Historical, Political, Philosophical and Mechanical Essays. The First, Containing an analysis of a General Map of the Middle British Colonies in America* (Philadelphia: B. Franklin and D. Hall, 1755). A fine study of Evans, with a convenient collection of reproductions of his maps, is Lawrence Henry Gipson, *Lewis Evans* (Philadelphia: Historical Society of Pennsylvania, 1939). A more recent study is Walter Klinefelter, "Lewis Evans and his Maps," *Transactions of the American Philosophical Society*, New Series, vol. 61, part 7 (Philadelphia, 1971).

12. "A Map of ye Mohegan Sachims Hereditary Country platted August the first 1705 per John Chandler Surveyor." London, Public Record Office, MPG 397. Enclosed with "Exhibits on behalf of the Mohegan Indians": C.O. 5/1272 ff. 342–3. See Clarence Winthrop Bowen, *The Boundary Disputes of Connecticut* (Boston: Osgood & Co. 1882), pp. 25–26. Lake Chargoggagoggmanchaugagogchaubunagungamaug is now over the boundary line in Massachusetts. In Indian times, as now, the lake was full of fish, and the passing Mohegan and Narragansett Indians continually fought with the local Nipmuck tribe over fishing rights. Finally, according to local tradition, they made a treaty that they embodied in the name of the lake. Chargoggagogg, I fish my side; manchaugagog, you fish your side; chaubunagungamaug, nobody fish middle. The name given on the Chandler survey is "Man hum squag [?]"; the present longer name is actually an early nineteenth-century combination of different names given to the lake. Possibly more accurate, if less romantic, is the etymology suggested by Harral B. Ayres, a local historian of nearby Webster, Massachusetts, which is based partly on correspondence with J. N. Hewitt of the Bureau of American Ethnology, Smithsonian Institution: "Englishmen at Manchaugagogg at the Boundary (or Neutral) Fishing Place." See Harral Ayres, *The Great Trail of New England* (Boston: Meador Publishing Co., 1940) p. 170; and a typewritten article by Mr. Ayres in the Webster, Massachusetts, Public Library.

13. Green (i.e., Braddock Mead) states on the face of the map that he used the surveys of Gardner and Kelloch but does not mention Douglass. Thomas Jefferys in 1768, however, gives Douglass as the basis for his "New England" in the table of contents of his *A General Topography*; see n. 36.

14. "A Map of the Colonies of Connecticut and Rhode Island, divided

into Counties & Townships . . . By Tho. Kitchin," *London Magazine* 27 (April 1758) opp. 168.

15. Thompson R. Harlow, "The Moses Park Map, 1766," *Connecticut Historical Society Bulletin* 28 (1963): 33–37; Edmund B. Thompson, *Maps of Connecticut Before the Year 1800* (Windham, Conn.: Hawthorn House, 1940), no. 16.

16. London. Public Record Office. C.O. 700 Rhode Island, 1.

17. *The Barrington-Bernard Correspondence and Illustrative Matter 1760–1770. Drawn from the "Papers of Sir Francis Bernard."* Eds. Edward Channing and Archibald Cary Coolidge (Cambridge, Mass.: Harvard University Press, 1912), pp. 51, 83.

18. Ibid., p. 70. Three maps of Maine made for Governor Bernard (dated 1764; unsigned) are in the Public Record Office: C.O. 700. Maine, 17, a route map from Penobscot to Quebec and return; C.O. 700. Maine, 18, thirteen townships laid out on the east side of Penobscot River; and C.O. 700. Maine, 19, seven townships laid out to the east of Mount Desert River.

19. "A Map of the Middle British Colonies in North America . . . By T. Pownall . . . Engraved by Jas. Turner in Philadelphia . . . Printed & Published for J. Almon, London. 1776," in *A Topographical Description* (London, 1776).

20. George E. Street, *Mount Desert: A History.* Ed. Samuel A. Eliot (Boston. Houghton Mifflin, 1926 [1st ed., 1905]), pp. 126–27. Mrs. Napier Higgins, *The Bernards of Abington and Nether Winchendon,* 2 vols. (London: Longmans, Green, 1903–4) 2:285–86.

21. *New Jersey Road Maps of the 18th Century,* ed. Howard C. Rice, Jr. (Princeton: Princeton University Library, 1970): John Dalley's "A Map of the Road from Trenton to Amboy" (1745), no. 1, and Azariah Dunham's "A Map of the Division line Between the Counties of Middlesex & Somerset" (1766), no. 2. Christopher Colles, *A Survey of the Roads of the United States of America, 1789,* ed. Walter W. Ristow (Cambridge, Mass.: Belknap Press of Harvard University Press, 1961). Miss Jeannette Black of the John Carter Brown Library has called my attention to R. Daniel's "A Map of ye English Empire in ye Continent of North America" (ca. 1679) as possibly being the earliest general map to show roads in the British colonies.

22. Library of Congress, Faden Collection no. 9 (manuscript).

23. British Museum, King's Topographical Collection cxx. 24.

24. Thomas Jefferys, *A General Topography* (London, 1768), no. 22.

25. Bancker's manuscript map: Vermont Historical Society, Montpelier, Vt. A copy of Roman's engraved map is in the John Carter Brown Library, Providence, R. I. See also David Allan Cobb, "Vermont Maps Prior to 1900: An Annotated Cartobibliography," *Vermont History* 39 (1971): 169–81.

26. "An Exact Mapp of New England and New York." Cotton Mather, *Magnalia Christi Americana: or, The Ecclesiastical History of New-England* . . . (London: Thomas Parkhurst, 1702), opposite p. A2.

27. William Douglass, M. D., *A Summary, Historical and Political, of the . . . Present State of the British Settlements in North America,* 2 vols. (Boston, 1749–51) 1:362; and Lloyd A. Brown, "Notes on the *Magnalia* Map," in Thomas James Holmes, *Cotton Mather: A Bibliography of His Works,* 3 vols. (Cambridge, Mass.: Harvard University Press, 1940), 2: 592–96.

28. "New England Observed and described by Captayn John Smith. The most remarqueable parts thus named. by the high and mighty Prince Charles, Prince of great Britian. Simon Pasaeus Sculpsit. Robert Clerke excudit. Printed by Geor: Low London." John Smith, *A Description of New England* (London: Robert Clerke, 1616). The nine different plates and states of the map are described in Joseph Sabin and others, *Bibliotheca Americana: A Dictionary of Books relating to America,* 29 vols. (New York: Bibliographical Society of America, 1892–1928) 22:223–24.

29. "The South part of New-England, as it is Planted this yeare, 1634." in William Wood, *New Englands Prospect* (London, 1634).

30. David Woodward, "The Foster Woodcut Map Controversy: a further examination of the evidence," *Imago Mundi* 21 (1967):52–61.

31. Tony Campbell, "New Light on the Jansson-Visscher Maps of New England," *Map Collectors' Series No. 24* (London: Map Collectors' Circle, 1965), has a detailed study of the changes in this series of maps.

32. Justin Winsor, *America,* 3:384.

33. Paris. Bibliothèque nationale. Ge. D. 11729 (1), (2), (3), (4), (four sheets).

34. Adam Smith's phrase describing Douglass (Sabin, *Americana,* 5:502); for Douglass's social, religious, scientific, political, and educational theories see Max Savelle, *The Seeds of Liberty* (Seattle: University of Washington Press, 1965).

35. William Douglass, *Summary,* 1:362. The *Summary* began to be published in fascicles by 1747; the first volume, printed by Rogers and Fowle in Boston, appeared in 1749, the second in 1752. It was reprinted in London by R. Baldwin in 1755 with a statement at the end of the table of contents in the first volume, "Place the Map to Face the Title of Vol. 1." Another reprint of the work, published in London by R. and J. Dodsley, appeared in 1760. None of the *Summary* volumes examined by or known to the writer has a copy of the rare Douglass map; some copies of the Dodsley reprint of 1760 have John Huske's "A New and Accurate Map of North America . . . London 1755." Douglass expected to

publish his map in the work; in vol. 2, p. 21, he writes, "I annex a correct map of the dominions of New-England, extended from 40 d. 30 m. to 44 d. 30 m. N. Lat.: and from 68 d. 50 m. to 74 d. 50 m. W. Longitude from London," to which he adds a footnote: "This map is founded upon a chorographical plan, composed from actual surveys of the lines or boundaries with the neighboring colonies (to Connecticut), and from . . . the records lodged in the secretaries office and townships records, with the writers perambulations: when this plan is printed, the author as a benefaction gives gratis, to every township and district, a copper plate copy; as the writer of the summary had impartially narrated the management of a late g——— which could not bear the light; to check the credit of the author the g——— endeavored (as shall be accounted for) to divert, impede, or defeat this publick generous-spirited amusement, but in vain."

36. After Dr. Douglass's death his executors sent the original draft to London, where it was engraved by R. W. Seale, according to a note at the top right of the map. Whether Jefferys pirated or purchased it is not known; in the table of contents of *A General Topography of North America* (London, 1768) he states, however, that the map of New England is based on that of Dr. Douglass. Green (Mead), who acknowledges other sources that he used, does not mention Douglass, although he follows the earlier map in exact detail for rivers, creeks, names of settlements, and boundary lines for the central area that they both have in common. The Jefferys-Green map covers a larger region, extending to the Hudson River in the west, Long Island to the south, and a greatly improved Maine coastline, based on the surveys ordered by Governor Shirley in 1754. The delineation of Lake Winnepesaukee is improved from Mitchell and Hazen's New Hampshire surveys, and Douglass's inset of Casco Bay is replaced by Fort Frederick on Lake Ontario and Boston Harbor.

37. The use of French sources in English cartography of Canada deserves further study. Helpful works on the French mapping of Canada before the middle of the eighteenth century are Don W. Thompson, *Men and Meridians* 1 (Ottawa: Queen's Printer and Controller of Stationery, 1966), with excellent bibliographies; Marcel Trudel, *An Atlas of New France* (Ottawa: Laval University Press, 1968); Nellis M. Crouse, *Contributions of the Canadian Jesuits to the Geographical Knowledge of New France, 1632–1675* (Ithaca: Cornell Public Printing Company, 1924); and Louis C. Karpinski, *Bibliography of the Printed Maps of Michigan, 1804–1880* (Lansing: Michigan Historical Commission, 1931), with numerous maps of the Great Lakes region reproduced. English use of French maps is illustrated in Edward Wells, *New Sett of Maps* (Oxford, 1700); Herman Moll, *Atlas Manuale* (London, 1709), with three strikingly different conceptions of the Great Lakes system in as many maps,

The World Described (London, 1709–20), and *Atlas Minor* (1729); Henry Popple, "British Empire in America with the French and Spanish Settlements adjacent" (London, 1733); and John Mitchell, "British Colonies in North America" (London, 1755).

38. Nathaniel N. Shipton, "General James Murray's Map of the St. Lawrence," *The Canadian Cartographer* 4 (1967):93–102.

39. Ibid., p. 100. Another account of the making of Murray's map is in Don W. Thompson, *Men and Meridians*, 1:97–98.

40. Edward Gaylord Bourne, "The Travels of Jonathan Carver," *American Historical Review* 11 (1906):288. Carver's maps are derived from surveys and sources not his own; indeed, the extent of his own authorship of them is questionable. An analysis of the historical information provided by his maps is in Emerson D. Fite and Archibald Freeman, *A Book of Old Maps* (Cambridge, Mass.: Harvard University Press, 1926), nos. 56, 64, 70. Bourne (p. 299) proves that the second part of Carver's *Travels,* describing the manners and customs of the Indians, "is essentially a compilation from La Hontan, Charlevoix, and Adair . . ." and suggests that the work was chiefly composed by its editor, Dr. John Coakley Lettsom, "a voluminous and facile writer and the charitable friend of Carver." A defense of Carver against some of Bourne's strictures is found in John T. Lee, *Captain Jonathan Carver: Additional Data* (Madison: State Historical Society of Wisconsin, 1913), pp. 87–123.

41. *Papers Relating to an Act of the Assembly* (New York: William Bradford, 1724) contains the map, regarded as the first map to be engraved in New York; a copy of the map and the twenty-four-page pamphlet is in the Public Record Office (M.P.G. 8, with C.O. 5/1092/82B). The work is the first edition of Colden's *The History of the Five Nations of Canada* (London, 1727), which has the same map from a new plate; a new and slightly revised map was made for the 1747 edition of Colden's work. Reproductions of the second and third states of the map are in Lloyd Arnold Brown, *Early Maps of the Ohio Valley* (Pittsburgh: University of Pittsburgh Press, 1959). pls. 13 and 15.

42. [A map of the Ohio country visited by John Patten during his captivity by the French, 1750–1751. Made by John Patten in Philadelphia, December, 1752]; (anonymous MS map in the Library of Congress). See Howard N. Eavenson, *Map Maker and Indian Traders* (Pittsburgh: University of Pittsburgh Press, 1949), with map reproduction at end.

43. [Map of the Ohio Company's lands on the Ohio River with the proposed location of a fort and settlement and with long explanatory legends. Signed "G. Mercer," ca. 1753]. London, Public Record Office.

44. [A sketch map portraying territory crossed by George Washington in 1753–1754 between Cumberland, Maryland, and Fort Le Boeuf, at Waterford, Pennsylvania, showing parts of the Potomac, Menongehela,

and forks of the Ohio and French Creek, 1754]. London, Public Record Office.

45. "Map of the Western parts of the Colony of Virginia, as far as the Mississipi," in George Washington, *The Journal of Major George Washington* (London: T. Jefferys, 1754). For Jefferys's use of Fry and Jefferson's MS map see Delf Norona, ed., "Joshua Fry's Report on the Back Settlements of Virginia (May 8, 1751)," *The Virginia Magazine of History and Biography* 56 (1948): 22–41.

46. Thomas Hutchins's dual map was engraved for William Smith, *A survey of that Part of the Country through which Colonel Bouquet Marched in 1764* (London, 1766): "A Map of the Country on the Ohio & Muskingum Rivers" (top), and "A Survey of that part of the Indian country through which Colonel Bouquet Marched in 1764" (bottom). Hutchins's large map (44″ x 36¾″), "A New Map of the Western Parts of Virginia . . . 1778," appeared in his *A Topographical Description of Virginia, Pennsylvania, Maryland, and North Carolina* (London, 1778). These and other selected maps, plans, and views of the trans-Appalachian region are reproduced, with an excellent general introduction and historical documentation for each map, in Brown, *Early Maps of the Ohio Valley*. Brown's *Early Maps* does not include, however, important maps in the Public Record Office that illustrate the Iroquois Indian country, western New York, and the Indian treaty boundary lines.

3. Charting the Coast

1. James Rosier, "A True Relation of the most prosperous voyage made this present yeere 1605, by Captaine George Weymouth," in Henry S. Burrage, *Early English and French Voyages, 1534–1608* (New York: Scribner's, 1906), pp. 361–62.

2. "Carta particolare della costa di Florida è di Virginia," *Dell' Arcano del Mare*, 3 vols. (Florence, 1646–47), vol. 3, bk. 6, no. 102. "Virginiae Item et Floridae Americae Provinciarum, nova Descriptio," Gerard Mercator, *Atlas . . . illustratus á Iudoco Hondio* (Amsterdam, 1606), no. 143. "Carta particolare della Virginia Vecchia è Nuoua," *Dell' Arcano del Mare*, vol. 3, bk. 6, no. 101.

3. "John Seller," *Dictionary of National Biography*, 21 vols. (Oxford: Oxford University Press, 1921–22) 17:1165–66; R. A. Skelton, *Decorative Printed Maps of the 15th to 18th Centuries* (London: Spring Books, 1965), p. 71; A. H. W. Robinson, *Marine Cartography in Great Britain* (Leicester: Leicester University Press, 1962); W. R. Chapin, "A Seventeenth-Century Chart Publisher: Being an Account of the Present Firm of Smith & Ebbs," *American Neptune* 8 (1948):1–24; Charles R. [a Proclamation directed against unauthorized reprints of J. Seller's *English Pilot* and *Sea Atlas*, 1671], British Museum C. 21 f. 1 (47).

4. *The English Pilot: The Fourth Book*. London, 1689 (Amsterdam: Theatrum Orbis Terrarum, 1967); this facsimile edition has an excellent introduction by Professor Coolie Verner, who records thirty-seven editions of *The Fourth Book* (1689–1794) and four other editions by different publishers, with a list of the first appearance of new maps introduced into the various editions. Verner also made a careful examination of the different plates and states of the Virginia and New York charts in *A Carto-Bibliographical Study of the English Pilot: The Fourth Book* (Charlottesville, University of Virginia Press, 1960).

5. "Nicholas Comberford and the 'Thames School'—Sea-Chart Makers of Seventeenth Century London." Professor Smith kindly sent me a revised and enlarged manuscript of his article, which he is preparing for press. See also an abstract of this paper in *Imago Mundi* 24 (1970):95.

6. W. P. Cumming, "The Earliest Permanent Settlement in Carolina: Nathaniel Batts and the Comberford Map," *American Historical Review* 45 (1939):82–89. See also: Cumming, *The Southeast in Early Maps*, pp. 144–45.

7. Cumming, "The Earliest Permanent Settlement . . . ," p. 87; see also Elizabeth G. McPherson, "Nathaniell Batts, Landholder on Pasquotank River, 1660," *North Carolina Historical Review* 43 (1966):72–73.

8. London, British Museum, Add. MS. 5414. 24.

9. London, Public Record Office. C.O. 700 Maryland 1; and P.R.O., M.P.G. 375.

10. See n. 4; Coolie Verner gives a list of additions and omissions in the various editions of *The English Pilot: The Fourth Book*.

11. Clara E. LeGear, "The New England Coasting Pilot of Cyprian Southack," *Imago Mundi* 11 (1954):137–44, analyzes the complexities in dating of *The New England Coasting Pilot* and lists Southack's cartographic contributions. The *Coasting Pilot* was reissued in the same size but as one map by W. Herbert and B. Sayer about 1755, and again reissued by Mount, Page, and Mount about 1775, though not in *The English Pilot*.

12. London, British Museum, Add. MS 5414, Roll 17; reprod. in A. B. Hulbert, *Crown Collection of Photographs of American Maps* 5 (Harrow, England: R. B. Fleming, 1908): no. 15.

13. London, Public Record Office, C.O. 700 Canada 1.

14. Lawrence C. Wroth, *Some American Contributions to the Art of Navigation 1519–1802* (Providence: John Carter Brown Library, 1947), p. 14.

15. London, Naval Library, Vv 2, vol. 2, no. 3, with the note "[?? Mapp Captain Thomas Smart in HMS. Squirrel in the Month of September 1718]."

16. Southack's hydrographic methods were certainly not in advance of

his time and already by 1753 John Green (Mead) had written with his acidulous pen "it does not appeare, however, that in making this chart he employed any instruments excepting the log and the Compass . . . without ever taking a single latitude": [John Green] *Explanation for the New Map of Nova Scotia and Cape Britain* (London: T. Jefferys, 1753), p. 5.

17. Wroth, *Some American Contributions*, p. 19.

18. Oxford, Bodleian Library, Rawlinson ms D. 908. fol. 207r: "A Large and aisect Drafe [exact draft] of the Sea Cost of No. Carolina." This map and Wimble's engraved map, "This Chart of his Majesties Province of North Carolina . . . 1738," are reproduced in William P. Cumming, "Wimble's Maps and the Colonial Cartography of the North Carolina Coast," *North Carolina Historical Review* 46 (1969):157–70.

19. William P. Cumming, "The Turbulent Life of Captain James Wimble," *North Carolina Historical Review*, 46 (1969):1–18.

20. G. R. Crone, *Maps and Their Makers* (London: Hutchinson University Library, 1953), p. 147.

21. G. R. Crone, "John Green. Notes on a Neglected Eighteenth Century Geographer and Cartographer," *Imago Mundi* 6 (1949):86.

22. See G. R. Crone, "Further notes on Bradock Mead, alias John Green, an eighteenth century cartographer," *Imago Mundi* 8 (1951):69–70, in which he corrects, on the basis of the two pieces of evidence discussed in this chapter, the impression given in his first article of the uneventfulness of Greene's life.

23. Thomas Mead was Lord Mayor of Dublin in 1757; no Braddock Mead is found in the list of alumni of Trinity College, Dublin, and Jefferys may have been mistaken: Crone, "Further notes on Bradock Mead," p. 69. Thomas Mead, a witness at Kimberley's trial in Dublin, was identified as a brother of Braddock: *The Case and Proceedings*, p. 22. "Edward Chambers" is evidently Ephraim Chambers (ca. 1680–1740), whose *Cyclopedia, or an Universal Dictionary of Arts and Sciences* was published in London in 1728 and revised and enlarged in 1738–39, 1741, and 1746; it influenced Samuel Johnson's method and ideas for his dictionary and, in a French translation, inspired Diderot's *Encyclopédie*: see "Ephraim Chambers," *Dictionary of National Biography*, 4:16–18. Edward Cave, a printer, originated and published the *Gentleman's Magazine*; Braddock Mead translated and revised D'Anville's maps in *A Description of the Empire of China and Chinese-Tartary* . . . . *From the French of P. J. B. Du Halde, Jesuit: with Notes Geographical, Historical, and Critical; and other Improvements, particularly in the Maps*, 2 vols. (London: Edward Cave, 1738–41). Thomas Astley was Cave's rival as publisher of *The London Magazine*; Mead was evidently the compiler and editor of *A New General Collection of Voyages and Travels: con-*sisting *Of the most Esteemed relations which have been hitherto published in any language* . . . . 4 vols. (London: Thomas Astley, 1745–47).

24. Library of Congress, Law Library, Law Trials (A & E) "Reading." "The whole Case and Proceedings in relation to Bridget Reading, an heiress. Containing an account of Kimberly's being sent to Ireland to bring over the said Bridget Reading, and of her pretended marriage to Braddock Mead. The information of Bridget Reading before Sir William Billers, against Braddock Mead, Daniel Kimberly, and Joseph Fisher, with the warrant of commitment, granted thereupon by the said Sir William Billers . . . to which is added, the tryal of the said Kimberly with his case, or last dying words, and an original letter sent by him to Mr. Reading, written some few days before his execution, and Mr. Reading's answer. London, Printed by and for R. Phillips; sold by E. Nott, H. Dodd, J. Jackson. 1730."

25. W. E. May, "The Surveying Commission of *Alborough*, 1728–1734," *The American Neptune* 21 (1961):260–78; Cumming, *Southeast*, pp. 51, 83, 194–95.

26. London, Naval Library, Vv 2, vol. 2, no. 39, "A Draught of Spirito Sancto with the coast adjacent, by Jas. Cook 1765"; vol. 2, no. 44, "A Draught of West Florida from Cape Blaze to Ibberville by Jas. Cook 1765" (with a note acknowledging his indebtedness to Mr. Pittman, Engineer of the 22nd Regt., to a gentleman of New Orleans for the River [Mississippi], and to a French pilot; and ending with a statement of his inability to secure a vessel to complete the survey though he had "solicited Captain [Rowland] Cotton to let me hire a vessel").

27. *Triangulation* is the measurement of the earth's surface by a network of triangles, usually with the aid of a theodolite; the surveying begins with an accurately measured baseline. Triangulation checks the accuracy of a survey by the measurements of the vertexes or angular points of the triangles; it also aids in correct geographical measurement of distances not easily measured by chain.

A *theodolite* is a surveying instrument used in triangulation for measuring horizontal and vertical angles. Its simplest form consists of a circle marked with degrees connected with a sighting mechanism and set on a tripod. In the eighteenth century its precision was improved by such additions as telescopic sights (first used by Jean Picard in 1669) and the vernier for measuring fractions of a degree. It was not until 1787 that Jesse Ramsden made for General William Roy a theodolite of exceptional accuracy that was used in the triangulation of the British Isles. A theodolite must be stationary and therefore used ashore; Captain John Gascoigne had a simple theodolite as part of his equipment in surveying Port Royal in 1728–29.

28. *Compass variation* is the difference between true north and the

magnetic north to which the compass needle points. A compass needle declines east or west from the true north except along the agonic line of no declination. Other factors, such as iron in the ship, may also cause variation of the compass; Captain James Cook complained of unexplained variation until he found the cause was an iron key that he had kept in the binnacle with the compasses.

29. Joseph Bernard, Marquis de Chabert de Cogolin, *Voyage Fait Par Ordre du Roi en 1750 et 1751, dans l'Amérique Septentrionale* (Paris, 1753), p. 2.

30. Instruments for determining the latitude or distance from the equator by observing the altitude of the Pole Star, such as the astrolabe and quadrant, had been in common use by mariners during the Middle Ages. In 1731 John Hadley made a reflecting octant that increased the possible angle of sighting to 90°. The modern sextant uses the same principle as Hadley's octant but can measure an arc of over 120° with many improvements and refinements.

31. *Lunar distance* is the angular distance from the moon to a given planet or star; the longitude of the observer can be determined by means of tables computed in nautical almanacs.

32. Harrison's marine chronometer was developed in response to an award of £20,000 offered by the British government's Board of Longitude in 1714 for a timekeeper that would keep time accurately within certain specified limits aboard ship. Harrison's first chronometer of 1736 was rejected as too large; in 1761 he submitted a watch 5″ across, with balances and with a temperature compensation and an escapement movement that he invented. This was tested on a voyage to the West Indies and return in 1762; it came well within the thirty-mile limit of error allowed by the act, and after bureaucratic delays Harrison was awarded the prize. For some years the Admiralty considered the watch too expensive to supply as equipment for naval vessels; the surveys for *The Atlantic Neptune* were made without the aid of chronometers.

33. London, Public Record Office, Colonial Office 700, Florida 3. De Brahm's equipment for his Florida coastal survey is given in Louis De Vorsey, Jr., "Hydrography: A Note on the Equipage of Eighteenth-Century Survey Vessels," *The Mariner's Mirror* 58 (1972):173–77.

34. Wilbur H. Siebert, "Bernard Romans," *Dictionary of American Biography*, 20 vols. (New York: Charles Scribner's Sons, 1928–35), 16:126. The best account of Roman's life, with a list of his writings and maps, is P. Lee Phillips, *Notes on the Life and Works of Bernard Romans* (Deland, Florida: Florida State Historical Society, 1924).

35. "A Plan of the Bay and Harbor of Boston surveyed agreeably to the Orders and Instructions of the Right Honourable the Lords Commissioners For Trade and Plantations to Samuel Holland Esqr. His Majesty's Surveyor General for the Northn. Districts of North America by Messrs. Wheeler and Grant (Drawn by Jas. Grant 1775)." For an account of Samuel Holland see Willis Chipman, "The Life and Times of Samuel Holland," Ontario Historical Society, *Papers and Records*, 21 (1924):11–90.

36. "A Collection of Charts of the Coasts of Newfoundland and Labradore, &c. . . . Drawn from Original Surveys by James Cook and Michael Lane . . ." London: T. Jefferys [ca. 1769, the date of the latest map]; also published as *The Newfoundland Pilot* (London: T. Jefferys, 1775); reprinted in facsimile, *James Cook Surveyor of Newfoundland . . .*, with an introductory essay by R. A. Skelton (San Francisco: David Magee, 1965).

37. "Old Ferrol Harbour" [Newfoundland; James Cook, 1764, H.M.S. Grenville] (Taunton, Hydrographic Department, 342 17i).

38. Stokes, *The Iconography of Manhattan Island*, 6 vols. (New York: Robert H. Dodd, 1915–28), 1:350.

39. *L'Esprit des Journaux* 3 (Paris, 1784):359–74. A brief comment on Des Barres's methods is given in G. N. D. Evans, "Hydrography: A Note on Eighteenth-Century Methods," *The Mariner's Mirror* 52 (1966):247–50.

4. THE CARTOGRAPHY OF CONFLICT

1. Thomas Paine, "Common Sense," in *The Writings of Thomas Paine*. Ed. Moncure D. Conway, 2 vols. (New York: Burt Franklin, 1902 [1969 reprint], 1:92.

2. London, Public Record Office, Colonial Office 700, Carolina 21, "A Draught of the Creek Nation. Taken in the Nation May 1757 by William Bonar." The William L. Clements Library has a very similar map, also dated May 1757, unsigned, with differences in spelling and location of settlements and without the bordering illustrations.

For the geographical importance of the information on William Bonar's map see Louis De Vorsey, "Early Maps as a Source in the Reconstruction of Southern Indian Landscapes," *Red, White, and Black: Symposium on Indians in the Old South. Southern Anthropological Society Proceedings*, no. 5., ed. Charles M. Hudson (Athens, Georgia, 1971), pp. 19–21.

3. John Richard Alden, *John Stuart and the Southern Colonial Frontier* (Ann Arbor: University of Michigan Press, 1944), p. 93. Fort Toulouse, "postes des Alibamons," had been built by the French in 1716 on the east bank of the Coosa River, four miles above its junction with the Tallapoosa in present-day Alabama.

4. Jefferys's plate, with a slightly different title, "A Prospective View of the Battle fought near Lake George, on the 8th of Septr. 1755 . . ." is,

like Blodget's, accompanied by *An Explanation . . . by Samuel Blodget, occasionally at the camp, when the battle was fought* (London, 1756); Jefferys's "View" is also found in his *A General Topography of North America . . .* (London, 1768), no. 37. See Justin Winsor, ed., *Narrative and Critical History of America,* 8 vols. (Boston and New York: Houghton Mifflin, 1887), 5:586, n. 4; and *American Printmaking: The First 150 Years* (New York: Museum of Graphic Art, 1969), nos. 22 and 23, pp. 22–23.

5. Coolie Verner, "Maps of the Yorktown Campaign 1780–1781," *Map Collectors' Series No. 18* (London: Map Collectors' Circle, 1965), nos. 5, 6.

6. "A Map of Halifax Survey'd by M. Harris," an inset in Jefferys's "A New Map of Nova Scotia with its Boundaries according to Mr. D'Anvile," published in a later issue with a changed title, "A Map of the South Part of Nova Scotia and it's Fishing Banks . . ." For a careful analysis of the map, its authorship and political implications, see John Carter Brown Library, *Annual Report* (Providence, 1958–59), pp. 46–50, and *Annual Report* (1960), pp. 34–5.

7. "A New Map of Nova Scotia and Cape Britain," in *Explanation for the New Map of Nova Scotia . . .* (London: T. Jefferys, 1755).

8. "Carte d'une Partie de l'Amerique Septentrionale . . . ," and "A Map Exhibiting a View of the English Rights relative to the Ancient Limits of Acadia"; the same maps are found in Thomas Jefferys, *A General Topography of North America* (London, 1768), nos. 22a, 22b.

9. "A New and Accurate Map of the English Empire in North America: Representing their Rightful Claim as confirm'd by Charters, and the formal Surrender of their Indian Friends; Likewise the Encroachments of the French, with the several Forts they have unjustly erected therein. By a Society of Anti-Gallicans . . . Decr. 1755 . . . . Wm. Herbert . . . . & Robt. Sayer [London]," with seven insets including "several Forts" of the French.

An equally virulent attack is found in a map published in London in 1752, a copy of which is in the Public Record Office, London: C. O. 700 M. P. America (N & S) no. 19: "North America Performed under the Patronage of Louis Duke of Orleans First Prince of the Blood by the Sieur D'Anville Greatly Improved by Mr. Bolton. Engraved by R. W. Seale MDCCLII . . . Gravelot delin. Walker sculps. Printed for John and Paul Knapton," in four sheets. A series of commentaries is printed on the map stating that boundary lines drawn by D'Anville have been excised as "a Romantic Presumption void of all authority"; "The French can have no shadow of Pretense to any part of this Country [south of the Saint Lawrence River] and their settlement about Lake Champlain is meer Depradation, contrary to Our Title." D'Anville's map "cost the late

Duke of Orleans 1000 pounds"; but French mapmakers are "advised to put their Louisiana farther West"; he made this map "for public consideration in order to avoid the future Calamities of war."

10. The author may be Ellis Huske, who lived in Boston and was Deputy Postmaster General of the colonies before his death in 1755, or it may be his son, John Huske, who lived in England and represented Maldon in the House of Commons. The writer has seen copies of the Huske map in William Douglass's *A Summary, Historical and Political . . .* (London, 1760), and in Thomas Pownall's *A Topographical Description . . . of North America* (London, 1776). Smaller maps are found in several magazines, with the same title, "A Map of the British and French Settlements in North America": *The General Magazine of Arts and Sciences* 1 (May 1755), engraved by Thomas Bowen; *The Universal Magazine of Knowledge and Pleasure* 17 (1755):145; *The Scots Magazine* 17 (1755):369, engraved by T. Phinn.

11. Intense interest on both sides of the Atlantic stimulated the commercial printing of maps, which are to be found in most major research libraries. Thomas Jefferys was the foremost engraver and publisher in England of maps of the French and Indian War. His work was characterized by excellence in engraving, in detail of execution (since many of his productions were large-scale maps of careful work by military surveyors), and was facilitated by exceptional access to military and naval sources of information. These sources were available to other mapmakers; but his position as geographer to the Prince of Wales (after 1760, to the king) may have made them more easily accessible for his use. The privileges and limitations of Jefferys's official position are commented on in R. A. Skelton, *James Cook, Surveyor of Newfoundland* (San Francisco: David Magee, 1965), p. 27, in the introductory essay to the facsimile editions of Jefferys's work; and in J. Brian Harley, "The Bankruptcy of Thomas Jefferys: an episode in the economic history of eighteenth century mapmaking," *Imago Mundi* 20 (1966):37. Most of the maps of Jefferys were originally published separately; but in 1768 he gathered many together for his *A General Topography of North America,* a fine atlas of one hundred maps that contains many of his productions relating to the conflict.

Jean Rocque's *A Set of Plans and Forts in America, Reduced from Actual Surveys* (London, 1763) contains thirty plans of fortifications and cities from New York to Quebec. Rocque, topographer to the king, died in 1762; he apparently prepared the atlas that was published by his widow, Mary Ann Rocque, whose name appears as publisher of a map of New York, the first map in the atlas. Many contemporary periodicals, such as *The Universal Magazine* and *The London Magazine,* published maps that frequently accompanied or illustrated reports or letters from the theater of war. From 1754 through 1760 *The London Magazine* published twenty-one maps, views, and plans of the North American scene; many of these were by Thomas Kitchin, who later became hydrographer to the king. Kitchin made many of the maps in the useful set found in Thomas Mante's *The History of the Late War in North-America* (London, 1772).

12. *The Journal of Major George Washington, sent by the Hon. Robert Dinwiddie . . . . to the Commandant of the French Forces on Ohio* (Williamsburg: William Hunter, 1754), p. 25. This booklet has no map; the reprint of the *Journal* in London by Thomas Jefferys later in 1754 includes a tracing of Washington's journey on a large map of the Ohio River Valley based on several sources including information sent by Fry and Jefferson that Jefferys did not use for their "Virginia" (1751, 1755); see p. 92, n. 45. Here may be mentioned a manuscript map made for or by Thomas Pownall in 1754: "The New laid out Roads by Order of ye Assembly of Pensylvania from Shippensburg to a Branch of Yoheogenni & from Alliquipis Gap to Will's Creek." This map, with an accompanying sheet of "Explanation," is among the Loudoun maps in the Henry E. Huntington Library, LO 530.

13. Public Record Office, M. P. G. 118, an enclosure to Governor Dinwiddie's letter of 29 January 1754. About 1755 Washington made or annotated other copies of this map with later additions from other sources: British Museum Add. MS. 15, 563. a., and a similar sketch in the Public Record Office reproduced in *The George Washington Atlas* (Washington: U. S. George Washington Bicentennial Commission, 1932), pl. 11. For location of major collections of manuscript maps of the French and Indian War see the bibliographical essay.

14. Original in the John Carter Brown Library: reproduced in Lloyd Arnold Brown, *Early Maps of the Ohio Valley* (Pittsburgh: University of Pittsburgh Press, 1959), pl. 30. Another copy, almost identical and probably also in Gist's own hand, is in the Henry E. Huntington Library, HM 1095.

15. The maps and plans are found in "A Journal of the expedition to North America in 1755, under General Braddock," British Museum King's MS. 212. 1–6. These maps were engraved and published by Thomas Jefferys in 1758 and later in his *A General Topography* (London, 1768, no. 31; Jefferys states in the table of contents that they were made by "an Aid de Camp to the General." This was Captain Orme: cf. Justin Winsor, *Narrative and Critical History of America,* 5:499–500, n. 3; and P. L. Phillips, *A List of Geographical Atlases* (Washington, D. C.: Government Printing Office, 1909), vol. 1, no. 1196, pls. 46–50. The maps also show the unwise distribution of the advance party of four hundred under Lieutenant-Colonel Thomas Gage, who survived his wounds to

become commander-in-chief at the beginning of the Revolution, and the plan of the field of battle on 9 July 1775. For other maps of Braddock's expedition see Justin Winsor, *Narrative and Critical History of America,* 5:498–580 and Brown, *Early Maps of the Ohio Valley,* pls. 26–30.

16. British Museum, King George III's Topographical Collection CXX1. 1. (MS).

17. William Parkman's *Diary* and a letter from Charles Lee to his sister (16 September 1758), both quoted in Justin Winsor, *Narrative and Critical History of America,* 5:597.

18. Thomas Mante, *The History of the Late War in North-America* (London, 1772).

19. Plan of Fort Ticonderoga on Lake Champlain drawn by Captain Thomas Abercrombie, 1758. B. M. King George III's Topographical Collection, CXX1. 9.1.c. The British Museum has fifteen manuscript maps and plans of forts in upper New York Province (1756–58) by Captain Abercrombie or attributed to him.

20. British Museum. King George III's Topographical Collection CXX1. 9.1 (MS). One inset on this map, apparently for New York, is blank.

21. The British Museum catalogs fifteen manuscripts of Louisbourg, including those made at the time of its first capture under General Pepperell in 1745; Amherst's collection adds nine more of especial interest, including captured French plans and a "Plan of the Mines and Demolition of the Fortifications of Louisbourg, 8th November, 1760" (Add. MS. 57703/3). The William L. Clements Library has "A survey of the city and fortress of Louisbourg . . . shewing the places the forces landed at, with the intrenchments and batterys . . . . The encampments of the several regiments . . . Survey'd by Wm: Bontein engineer . . . under Colo. Bastide chief engineer . . . Copy Richd. Coombs," among the Germain Papers. See Christian Brun, *Guide to the Manuscript Maps in the William L. Clements Library* (Ann Arbor: University of Michigan, 1959), no. 45.

Among the printed maps are "A Plan of the harbour and town of Louisbourg in the island of Cape Breton, drawn on the spot," in *An Authentic Account of the Reduction of Louisbourg in June and July 1758. By a spectator* (London, 1758); "A Plan of the city & fortifications of Louisbourg," in *The London Magazine* 27 (August 1758): opposite p. 379; and "Plan of the city and fortifications of Louisbourg with the attacks," in John Rocque, *A Set of Plans* (London, 1763), no. 3.

22. British Museum, Add MS. 57712/6 (William Brasier; Plan of Fort George, 17 July 1759); this is a finely drawn map. Many maps of Fort William Henry and Fort Edward on the Hudson River, both strongly fortified during the years after 1755, are in the Amherst and King George III collections in the British Museum.

23. Jefferys, *A General Topography*, no. 12. Jefferys engraved another plan of the siege of Quebec, "An Authentic Plan of the River St. Laurence from Sillery to the Fall of Montmorenci," for his *The Natural and Civil History of the French Dominions in North and South America* (London, 1760), pl. 1, opposite p. 131, which was also included, with a slightly different imprint, in *A General Topography*, no. 10. Many contemporary manuscript maps of the operations around Quebec are found in the British Museum, Library of Congress, W. L. Clements Library, and elsewhere. Noteworthy is the large (27½″ x 80″) manuscript "Plan of Quebec" by Debbing, Holland, and Des Barres in the Library of Congress, with an attached descriptive text of the defense and attack of Quebec.

24. British Museum, Add. MS. 57708/4 (Oswego, Brasier, surveyor and draftsman); ibid., Add. MS. 57711/13 (Sowers and Brasier, Oswego); ibid., Add. MS. 57708/6 (Fort Ontario, Sowers and Brasier); ibid., Add. MS. 57708/8 (southern shore of Lake Ontario, the route from Oswego to Niagara, Sowers); ibid., Add. MS. 57708/9 (Fort Niagara, Lieutenant George Demler); ibid., Add. MS. 57708/10 (Plan of the Fort and Attack of Niagara, Quartermaster Jones of the 60th Regiment); other maps supplement these. Comparatively few maps of large areas are among the Amherst maps, but two are unusual: Add. MS. 57712/15 (Draft of a Scout over the Mountains West of Lake Champlain, I. Davies fecit, 1759); and Add. MS. 57707/1 (Sketch of the River Saint Lawrence from Ontario to Montreal, by an Onondaga Indian, drawn by Guy Johnson, 1759). Francis Pfister, a lieutenant of the First Battalion of the Royal American Regiment and later captain, made a series of plans during 1760 of the city of Albany, the forts of Oswego, Ontario, Ticonderoga, Crown Point, Edward, and of Montreal; these colored and finely executed plans for Amherst are in King George III's Topographical Collection and R.U.S.I. in the British Museum as well as his maps of larger areas.

Jean Rocque, *A Set of Plans*, contains maps of Fort Ontario (no. 30) and of Fort Niagara (no. 28).

25. British Museum, Add. MS. 57713/4; a similar unsigned version with a slightly different title but of the same year is in the British Museum, King George III's Topographical Collection, CXX1. 51. William Brasier (Brassier) was a surveyor and draftsman for Captain Thomas Sowers, chief engineer, and later for Captain Harry Gordon, chief engineer. His own maps are dated between 1759 and 1762; but he was evidently employed to copy maps of Lieutenant George Demler, Lieutenant Philip Pittman, and others, until 1766 and perhaps into the early 1770s. See Brun, *Guide*, nos. 372, 612, 682, 684, etc. A careful study will probably identify many unsigned maps of the period as copies by Brasier.

26. Jefferys, *The American Atlas*, no. 18, and Sayer and Bennett, *The American Military Atlas*, no. 6. Manuscript drafts of this map by Brasier are in the Library of Congress (Faden Collection, no. 19) and in the William L. Clements Library (Brun, *Guide*, no. 446).

27. Louis De Vorsey, *De Brahm's Report* . . . (Columbia: University of South Carolina, 1971) reproduces seven plans of forts in South Carolina and Georgia by De Brahm; see also Cumming, *Southeast*, pp. 246–50. Manuscript maps by De Brahm are in the British Museum, Public Record Office, Harvard, and other libraries.

28. British Museum Add. MS. 14036.e; "A Map of the Cherokee Country." See also: Alden, *John Stuart*, pp. 113, 365, and Cumming, *Southeast*, p. 231.

29. British Museum, Add. MS. 57714/16; "Sketch of the Cherokee Country and march of the troops under command of Lieutenant Col. Grant to the middle and back settlements. 1761."

30. Henry Timberlake, *The Memoirs of Lieut. Henry Timberlake* (London, 1765), opposite p. 1; the map is also in Jefferys, *A General Topography*, no. 38.

31. *A List of the General and Field Officers as They Rank in the Army. A List of the Officers in the Several Regiments . . . as They Rank in Each Corps* [London, 1754–1868]: Ratzer's commission (1757 volume); Captain Lieutenant and Captain, 14 August 1773 (1774 volume); Captain (1776 volume); date of commission as captain changed to 13 November 1772 (1777 volume; earlier date an error?); "Rank in Regt., Captain," and rank in army, "Major 19 Mar. 1783" (1783 volume). Perhaps this last rank had something to do with demobilization, or with higher ranking in the colonial system. He is not listed in the army list for 1784 under any category.

32. *Journals of the Assembly of Jamaica*, 14 vols. (London, 1811–29), 6:477 (10 December 1773; £100 for plans); 6:483 (17 December 1773; £20 to "Benjamin [sic] Ratzer, assistant engineer"); 6:517 (1 November 1774), 550–51 (9 December 1774), 556 (14 December 1774), 695 (19 December 1774), Ratzer, who had been ill of a severe fever for eighteen days, requests more than the £150 of pay for services, as his house rent alone has been £65; the assembly votes termination of his service as sub-engineer, and confirms pay of £150; 7:32 (5 December 1777), 98 (23 November 1778) Gilbert Waugh of Port-Royal petitions payment of £268.10s. for use of a wherry and five Negroes for 358 days by Captain Bernard Ratzer of Fort Augusta and his predecessor; assembly orders investigation of facts and of authority for employing the wherries. I am indebted to Dr. David J. Buisseret of the Department of History, University of the West Indies, and to Miss Jeannette Black of the John Carter Brown Library for their suggestions and aid in tracing Ratzer's career.

33. Unlike some of his predecessors as chief engineer, his father James, Captain Harry Gordon, and others, John Montresor made notable con-

tributions to North American cartography. Montresor, who incurred General Clinton's displeasure several times, was superseded in 1778 and retired to England. For his own account of his experiences and activities in America from 1757 to 1778, see *The Montresor Journals*, ed. G. D. Scull (New York, Printed for the [New-York Historical] Society, 1882); some Montresor letters are in Gideon D. Scull, *The Evelyns in America . . . 1608–1805* (Oxford: Parker and Co., privately printed, 1881); J. C. Webster, "Presidential Address: Life of John Montresor," *Transactions of the Royal Society of Canada,* ser. 3, vol. 22, sect. 2 (May 1928) gathers other biographical information. A study of Montresor's cartographical work, much needed, is in progress by Douglas W. Marshall of the William L. Clements Library.

The name of William Brazier (also spelled Brasier and Brassier on his drafts) does not appear on American manuscript maps after 1766; on 19 August 1783 his widow, Catharine Brazier, in a petition to Sir Guy Carleton, requests continuance of her quarterly allowance previous to sailing for Nova Scotia from New York. Brazier is referred to as draftsman to the commanding engineer and chief draftsman to the Board of Ordnance. See *Report on American Manuscripts in the Royal Institution of Great Britain,* 4 vols. (London: Historical Manuscripts Commission, 1904–9), 3:301; 4:317.

34. The cartographical record of the revolutionary war is extensive. The chief printed sources for investigation include the following. Thomas Jefferys, *The American Atlas* (London: R. Sayer and J. Bennett, 1775), was followed by several later editions of the same work. Sayer and Bennett also published the *American Military Pocket Atlas* (London, 1776), known as the "Holster Atlas" because it was designed for use by mounted British officers. Its maps, like those in *The American Atlas,* are chiefly of large areas, based on information gathered before 1775. William Faden, *The North American Atlas* (London, 1777), contains many maps by military surveyors like Sauthier and Blaskowitz showing battles and important military areas after the outbreak of the war. William Faden, *Atlas of Battles of the American Revolution* (London, ca. 1793; reprint from original plates: New York, Bartlett and Welford, 1845?), is a collection of plans and maps of battles engraved by Faden during and after the war. *The Atlantic Neptune* (London, ca. 1774–81) has many charts with land areas adjacent to the coast, surveyed with accuracy and detail.

The London Magazine, Gentleman's Magazine, Political Magazine, and *Universal Magazine* have numerous maps for the period; they are sometimes simplified from larger maps by omission of irrelevant detail, with insertion of pertinent troop movements or battle information. Not to be overlooked, even though they may have been published after the conclusion of the war, are maps included in the journals and memoirs of the participants; among these are Charles Stedman, who served under Sir W. Howe, Sir H. Clinton, and the Marquis Cornwallis, *The History of the Origin, Progress, and Termination of the American War,* 2 vols. (London, 1794), with three maps and twelve plans, and Lieutenant-Colonel John G. Simcoe, *A Journal of the Operations of The Queen's Rangers* (Exeter, 1787), with ten order-of-battle plans, some of which are based on sketches "taken on the spot." For location of major collections of manuscript maps of the revolutionary war see the bibliographical essay.

35. Faden, *North American Atlas,* nos. 13–14.

36. Ibid., no. 13.

37. Jefferys, *American Atlas,* no. 20 (with 1775 imprint).

38. Ibid., nos. 23–24.

39. Cf. n. 25, with reference to William Brasier.

40. Among the most productive mapmakers for the British after 1775 are Major John André, Deputy Surveyor Charles Blaskowitz, Captain Edward Barron, Commanding Engineer Abraham D'Aubant, Major Patrick Ferguson, Captain Edward Fage, Lieutenant John Hills, Lieutenant (later Sir) Thomas Page, Lieutenant-Colonel Lord Francis Rawdon, Lieutenant George Sproule. Some of these, such as the gallant and unfortunate Major André, did not survive the war. Others went on to greater fame like Lord Rawdon, who as ensign was a protégé of Earl Percy in Boston, became a lieutenant-colonel in command of British troops in South Carolina when Cornwallis left for Virginia in 1781, and still later as Lord Hastings became governor-general of India.

41. Scale regulations issued by the Corps of Engineers were, for a general map of a coast or island, 1″ to 1600′; a town and environs, 1″ to 800′; a town or settlement, 1″ to 400′; a fort or battery, 1″ to 200′; a magazine or building, 1″ to 10′; a drawbridge or gun carriage, 1″ to 5′: Whitworth Porter, *History of the Corps of Royal Engineers,* 3 vols. (London: Longmans, Green, and Co., 1889–1915), 1:150. There is a need for further investigation of what cartographic training was given to officer candidates and what instruments were provided engineers in the field through an examination of contemporary journals, audit accounts, and other sources.

42. See Appendix B for a fuller analysis of the map; see also William P. and Elizabeth C. Cumming, "The Treasure of Alnwick Castle," *American Heritage* 20, no. 5 (August 1969):22–23, 99–101; reproduction, p. 29.

43. Ibid., p. 28.

44. General William Howe planned to appoint Barron as chief engineer in 1777, but Barron was taken prisoner; by an arrangement made by M. G. Massey with the American General Artemas Ward in January

1778 Barron was returned in an exchange of prisoners: Howe to Massey, 17 March 1777; Massey to Howe, 2 January 1778, and 12 January 1778. *Report on American Manuscripts in the Royal Institution of Great Britain,* 4 vols. (London: Historical Manuscripts Commission, 1904–9), 1:178. For maps Barron made for Lord Percy in Nova Scotia in 1779 and of Chester, England, in 1786 see Appendix B.

45. Letter by Captain John Chester dated 22 July 1775, in Henry Steele Commager and Richard B. Morris, eds., *The Spirit of 'Seventy-Six,* 2 vols. (Indianapolis and New York: Bobbs-Merrill, 1958), 1:125.

46. Lieutenant Thomas Page made a detailed "A Plan of the Action at Bunkers Hill," on surveys by Montresor, which is in the Library of Congress Faden Collection, no. 22, engraved, without Page's name, for C. Stedman's *History . . . of the American War* (London, 1794), 1: opposite 127. Page also produced a well-known "A Plan of the Town of Boston with the Intrenchments &c. of His Majestys Forces in 1775," engraved in 1777 by William Faden, *Atlas of Battles of the American Revolution* (London, ca. 1793); the original manuscript, in the Library of Congress Faden Collection, no. 32, is reproduced in E. D. Fite and A. Freeman, *A Book of Old Maps* (Cambridge, Mass.: Harvard University Press, 1926), no. 62. Page also made another map, "Boston, its Environs and Harbour, with the Rebel Works . . . Engraved & Publish'd by Wm. Faden . . . 1777"; for the different issues of this map see Henry Stevens and Roland Tree, "Comparative Cartography," *Map Collectors' Series No. 39* (London: Map Collectors' Circle, 1967), pp. 5–6.

Montresor made several copies of a beautifully executed, colored "Draught of the Towns of Boston and Charles Town and the Circumjacent Country" during the siege of Boston, manuscript copies of which are in the Percy Collection at Alnwick Castle, W. L. Clements Library, and the Library of Congress.

47. "The Seat of Action between the British and American Forces; or an Authentic Plan of the Western Part of Long Island . . . from the Surveys of Major Holland . . . R. Sayer & J. Bennett, 1776." A manuscript "Plan of the Attack [of] the Rebels on Long Island by an Officer of the Army" is in the Library of Congress Faden Collection, no. 56. The events on Long Island and subsequent engagements are given also in "A Plan of New York Island with part of Long Island . . . Engraved and Publish'd . . . 1776 by Wm. Faden"; five different issues of this are examined in Stevens and Tree, "Comparative Cartography," no. 41.

48. Faden's engravings of Sauthier's maps of the New York campaign are found in his atlases and also as separate sheets.

"A Plan of the Operations . . . in New York and East New Jersey . . . on the White Plains . . . By Claude Joseph Sauthier: Engraved by Wm. Faden, 1777." Faden, *Atlas of Battles of the American Revolution,* no.

34. The original draft for this map, elaborately detailed and in color but without Sauthier's name, is in Library of Congress Faden Collection, no. 58.

Stokes, *Iconography,* 1:355, has the comment on Sauthier's map, "A Topographical Map of the Northn. Part of New York Island . . . Survey'd immediately after [the battle] . . . 16th. Novr. 1776 . . . by Order of his Lordship [Earl Percy]. By Claude Joseph Sauthier . . . London . . . 1777. by Wm. Faden," *North American Atlas* (London, 1777), no. 18. Sauthier's original manuscript draft of this map, on which he states it was surveyed on the day of the battle by Percy's order, is in the Library of Congress Faden Collection, no. 61. The Faden Collection has other manuscript maps by Sauthier: no. 95, "A Plan of Fort George at the City of New York" (ca. 1772), and no. 58, not signed but in Sauthier's style, "A Plan of the Operations of the King's Army . . . in New York and East New Jersey . . . the Engagement on the White Plains the 25th of October [1776]." The Faden engraving of Sauthier's map of New York is reproduced in W. P. and E. C. Cumming, "The Treasure of Alnwick Castle," p. 31.

Two maps of the Long Island and Fort Washington campaigns with lists of references are in the Clinton Papers (W. L. Clements Library; Brun, *Guide,* nos. 383, 384). They are dated 1777 and are signed by Ferdinand Joseph Sebastian De Brahm, a nephew by his elder brother of Surveyor General De Brahm. Ferdinand, a trained engineer and cartographer, became a Patriot who received on 11 February 1778 a commission as "Major Engineer" in the Continental Army; apparently his activities put his uncle in difficulties with the British authorities, who confused Ferdinand with the loyalist surveyor general: De Vorsey, *De Brahm's Report,* pp. 53–54.

49. Blaskowitz's survey of Rhode Island, part of a general survey of the coast from the Bay of Fundy to Rhode Island, was announced in *The Massachusetts Gazette: and Boston Weekly News-Letter,* Thursday, 5 May 1774; reprinted in *The Magazine of American History* 8 (December 1882; New York and Chicago: A. S. Barnes & Co.): 797–800. The map is found in Faden, *North American Atlas,* no. 11.

50. See Appendix B: Percy Collection.

51. Faden, *Atlas of Battles of the American Revolution,* no. 35: drawn by Mr. Medcalfe and published in 1780. Other general maps of the campaign are "Map of Hudson's River," *Gentleman's Magazine* 48 (January 1778): opposite 41, and "Part of the Counties of Charlotte and Albany, . . . being the Seat of War between the King's Forces . . . and the Rebel Army. By Thos. Kitchin," *London Magazine* 47 (February 1778):50. General John Burgoyne himself published several plans of battle in a defense of his conduct of the campaign after his return to England, in his *A State*

of the Expedition from Canada as laid before the House of Commons . . . with a Collection of Authentic Documents (London, 1780).

52. Faden, *Atlas of Battles of the American Revolution,* no. 30.

53. Ibid., no. 33.

54. Ibid., no. 17.

55. Ibid., no. 4; for other maps connected with the various attacks around West Point see Winsor, *America,* 6:557–58, and Brun, *Guide,* pp. 106–10.

56. "Plan of the Operations of General Washington . . . New Jersey, from the 26th. of December 1776. to the 3d. January 1777," Faden, *North American Atlas,* no. 21.

57. "Sketch of the Surprise of German Town, by the American Forces . . . October 4th. 1777; by J. Hills," Faden, *Atlas of Battles of the American Revolution,* no. 12.

58. See Appendix B: Percy Collection.

59. Faden, *Atlas of Battles of the American Revolution,* no. 23.

60. Phillips, *Atlases,* 1339 and 2143: the two collections of folio manuscript plans are almost duplicates, with twelve of the twenty plans in each signed by John Hills. Other manuscript maps by Hills are in the British Museum and the William L. Clements Library. Hills was assistant engineer to Clinton as early as 1778 (Brun, *Guide,* no. 511, "A Sketch of Haddonfield . . . March, 1778"); in the British army *List of General and Field Officers* for 1781–84 his rank is given as "Second Lieut. 23d. Regt. foot." In his "A Plan of Yorktown and Gloucester . . . 1781 . . . Printed Oct. 7th. 1785" (Faden, *Atlas of Battles of the American Revolution,* no. 15), he is "late Lieutt. in 23d. Regt. & Asst. Engrr.". Probably he settled in Philadelphia after the close of the war; in 1786 John Hills, "successor to Mr. Vancouver," advertises his services as "Land surveyor, architect and draftsman in Market-street, next door to Mr. Rice's Book-store, nine doors east of Third-street": *Philadelphia Herald, and General Advertiser* Monday, 5 July 1786 (Philadelphia, 1786, vol. 3, no. 47). In the *New Jersey Gazette,* Monday, 11 October 1784, appears a proposal for a map of New Jersey by John Hills, surveyor, architect, and draftsman, who also offers his services as land appraiser, with his address in Princeton, N. J. In the *Pennsylvania Packet, and Daily Advertiser,* Saturday, 27 February 1790, he requests subscriptions for a $2.00 detailed plan of Philadelphia. This is probably Hills's "Plan of the City of Philadelphia and its Environs . . . May 30th. 1796 . . . London, Engraved by John Cooke. Philadelphia, J. Hills, 1797." A later survey of ten miles from the center hydrant of the city resulted in "Plan of the City of Philadelphia and Environs. Surveyed by J. Hills, in the summers of 1801–7. Philadelphia, 1808," which is called by Justin Winsor "the best map of the suburbs of Philadelphia near this time" (*America,* 7:333). Hills evidently called on his earlier

map engraved by Faden in making "A Map of the Stage Routes between the Cities of New-York, Baltimore and Parts Adjacent. To Which is Added . . . the Operations of the British Army, from their Landing at Elk River in 1777 to their Embarkation at Nevisink in 1778. By John Hills, Engr. & Geographer. Philadelphia, E. Savage, 1800."

Hills, with his extensive surveys of routes for Clinton's military use and his detailed, colored, topographical maps of towns and counties of New Jersey, which sometimes included individual houses and their owners, was one of the prolific and able mapmakers of the revolutionary era. Some additional references to Hills may be found in the biographical card index in the Geography and Map Division, Library of Congress.

61. "A Plan of the Attack of Fort Sulivan . . . on the 28th of June 1776 . . . Engraved by Wm. Faden. 1776," Faden, *North American Atlas,* nos. 29, 30, appears to be based on a map by Lieutenant-Colonel Thomas James; an identical manuscript map by him is among the Clinton Papers in the W. L. Clements Library (C. Brun, *Guide,* no. 623). "A Plan of . . . the attack on Fort Sulivan . . . in 1776," *Political Magazine* 1 (1780): opposite 115, is reproduced in Winsor, *America,* 6:170.

62. An especially handsome map of the lower reaches of the Savannah River is "Sketch of the Northern Frontiers of Georgia, extending from the Mouth of the River Savannah to the Town of Augusta. By Archibald Campbell, Lieutt. Coll. 71st Regt. . . . May 1st. 1780 by Wm Faden": Jefferys, *American Atlas,* no. 25½. A colored topographical manuscript shows Campbell's attack: "Plan of the decent [*sic*] and action of the 29th. Decr. 1778 . . . under the command of Lt. Colol Campbell of the 71st Regt. foot. John Wilson, asst. engineer." (William L. Clements Library; C. Brun, *Guide,* no. 636). The French and American counterattack is found in "Plan of the Siege of Savannah . . . on the 9th. of October 1779 . . . [1784]," Faden *Atlas of Battles of the American Revolution,* no. 3; and the "Plan of the Siege of Savannah . . . surveyed by John Wilson," *Atlantic Neptune* (London, 1775–80), no. 16. A number of manuscript maps related to the siege are in the Library of Congress (Faden Collection no. 42), W. L. Clements Library (C. Brun, *Guide,* nos. 637–41), and British Museum, Add. MS. 57716/3). The R.U.S.I. map is by James Moncrief, commanding engineer of the 71st Reg. of Foot.

63. Phillips, *Atlases,* vol. 1, no. 1205, p. 654, chart 34.

64. Faden, *Atlas of Battles of the American Revolution,* no. 10.

65. Ibid., no. 8.

66. Banastre Tarleton, *A History of the Campaigns of 1780 and 1781, in the Southern Provinces of North America* (London: T. Cadell, 1787), opposite p. 1. Several of Faden's maps relating to the Southern campaign are reproduced in this volume and in Stedman's *History . . . of the American War* (London, 1794).

67. Faden, *Atlas of Battles of the American Revolution,* no. 7; Tarleton, *History,* p. 108; Stedman, *History,* 2: opposite 342. These are all identical with a manuscript map of the same title in the Clinton Papers (C. Brun, *Guide,* no. 592).

68. W. Faden, *Atlas of Battles of the American Revolution,* no. 9; reproduced in J. Winsor, *America,* 6:543. C. Stedman, *History . . . of the American War* (London, 1794) 2: opposite 358, uses Faden's plate with minor changes.

69. John R. Alden, *A History of the American Revolution* (New York: Knopf, 1969), p. 465.

70. British Museum, Add. MS. 57715/11: "March of the army under Lieutenant General Earl Cornwallis in Virginia from the junction of Petersburg on the 12th of May till their arrival at Portsmouth on 12th July 1781."

71. John G. Simcoe, *A Journal of the Operations of the Queen's Rangers* (Exeter: for the author, 1787); manuscript maps in the Simcoe Papers, Henry E. Huntington Library.

72. Coolie Verner, "Maps of the Yorktown Campaign," *Map Collectors' Series No. 18* (London: Map Collectors' Circle, 1963).

73. British Museum, Add. MS. 57715/14–17 [four positions of the British and French fleets as the latter comes out of Chesapeake Bay]. Verner, "Yorktown," lists several printed and manuscript maps and charts of fleet and naval actions during the Yorktown campaign.

74. Neither British army activity nor British mapmaking stopped immediately. Charleston, South Carolina, remained under military rule until 14 December 1782. In December 1781 the safety of the garrison had become a matter of concern to the officers, leading to the construction of new fortifications: Hampstead Hill, the New Horn Work, and the New Bastion. These are marked on a map endorsed "Charles Town Neck. Exhibiting the Plan of the Town and all the Fortifications in December.

1781," British Museum, Add. MS. 57715/18. The map also shows the defenses built by General Lincoln in 1780.

75. Sauthier made more than one copy of several of these maps; for their present location see Cumming, *Southeast,* pp. 86–87, 239–45.

76. Public Record Office, A. O. 12/24.

77. Details of his Strasbourg background have been furnished through the kindness of M. P. Schertzer of the Strasbourg Archives and of M. Louis Schlaefli, librarian of the Grand Séminaire.

78. Franz Grenacher, "Current Knowledge of Alsatian Cartography," *Imago Mundi* 18 (1964):69–70. I am indebted to Dr. Grenacher for helpful suggestions in tracing the early career of Sauthier.

79. Strasbourg, Bibliothèque du Grand Séminaire, MS. 316.

80. Alonzo Thomas Dill, *Governor Tryon and His Palace* (Chapel Hill: University of North Carolina Press, 1955), pp. 119, 126.

81. P. R. O., A. O. 12/26; John T. Faris, *The Romance of the Boundaries* (New York. Harper & Brothers, 1926), p. 50, notes that Sauthier and John Collins, deputy surveyor general for Quebec, ran the line westward from Lake Champlain to within ten miles of the Saint Lawrence, but that Collins, by agreement between the governors of New York and Quebec, ran the last ten miles for £100 in September 1772.

82. Norbury is on the south branch of the Lemoille or Scodoqua River, which flows into Lake Champlain; it was in an area of dispute between New York and New Hampshire.

83. Faden Collection, no. 61; later engraved and published by William Faden in his *North American Atlas* (London, 1777), no. 18.

84. Public Record Office, W. O. 12/24.

85. See Peter J. Guthorn, *American Maps and Map Makers of the Revolution* (Monmouth Beach, N. J.: P. Freneau Press, 1966), and his *British Maps of the American Revolution* (Monmouth Beach, N. J.: P. Freneau Press, 1972).

A Bibliographical Essay

This essay is a guide to the major general sources of information concerning the eighteenth-century British cartography of North America. It lists the chief bibliographical works in the field, both general and regional, but does not include specific studies of individual maps and mapmakers. Since its purpose is to facilitate further investigation, it includes also a few of the major atlases published in England during the eighteenth century that show both the increase of geographical knowledge and the development of cartographic skills; the most useful published collections of reproductions; and some of the chief collections of colonial maps in British and North American libraries.

General Works
Map Lists and Catalogues

A basic list of American maps is Philip L. Phillips, *A List of Maps of America in the Library of Congress Preceded by a List of Works Relating to Cartography*, (Washington: Government Printing Office, 1901), with over 15,000 maps listed alphabetically by area. A standard reference work on atlases is *A List of Geographical Atlases in the Library of Congress*, (Washington, D.C.: Government Printing Office), vols. 1–4 compiled by Philip L. Phillips, 1909–20, and vols. 5–6 compiled by Clara E. Le Gear, 1958, 1963, with vol. 7 at present in press. Another general list of printed maps is *The British Museum Catalogue of Printed Maps, Charts and Plans,* 15 vols. (London: Trustees of the British Museum, 1967). It is supplemented by the *Catalogue of the Manuscript Maps, Charts, and Plans, and of the Topographical Drawings in the British Museum 3* (London: printed by order of the Trustees, 1844–61), which has the

maps and plans in the British Museum at the time of its publication, including the rich collections of George III presented to the museum by George IV in 1823 and later. *The Catalogue of Maps, Plans, and Charts in the Library of the Colonial Office,* 2 vols. (London: Public Record Office, 1885), will be replaced eventually by an enlarged general catalogue of American maps in the Public Record Office that is being compiled by Mr. P. A. Penfold and Mr. H. N. Blakiston from the card catalogue in the PRO.

One of the best general sources of information, with many facsimiles of contemporary maps, is Justin Winsor, *Narrative and Critical History of America,* 8 vols. (Boston: 1884–89), which includes occasional essays on the cartography of various regions and periods. More specialized but useful are I. N. Phelps Stokes and Daniel C. Haskell, *American Historical Prints* (New York: New York Public Library, 1932); and James C. Wheat and Christian Brun, *Maps and Charts Published in America before 1800,* (New Haven: Yale University Press, 1969). John Carter Brown Library, *Annual Reports* (Providence: John Carter Brown Library, 1911–   ), has numerous valuable short essays, chiefly by its erudite librarian, Lawrence C. Wroth, on eighteenth-century maps and charts acquired by the library through the years. The library has also published from time to time annotated catalogues of its exhibits that contain reproductions of maps. Henry Stevens and Roland Tree, "Comparative Cartography," *The Map Collectors' Series No. 39* (London: Map Collectors' Circle, 1967), examines the different plates and states of some one hundred maps of colonial America.

There are several periodicals that may be consulted for pertinent articles on special subjects; *Imago Mundi: A Review of Early Cartography* 1– (1935–   ); *Map Collectors' Series* 1– (1965–   ); *The Canadian Cartographer* 1– (1964–   ). The first four volumes of the last (1964–67) were published with the title *The Cartographer*.

Reproductions of Maps

A major collection of reproductions of British colonial maps is Archer B. Hulbert, *The Crown Collection of Photographs of American Maps* (Cleveland, Ohio [also Harrow and London], privately printed, 1904–16); Series I, 5 vols. British Museum [1904–8]. Series II, 5 vols. British Museum [1909–12]; Series III, Colonial Office Library, Public Record Office [1914–16]; Series IV [blueprints of Western maps] (Colorado Springs: Steward Commission on Western History [1925–28]).

The "Karpinski Series of Reproductions" are photographs of American maps that Professor Louis C. Karpinski made in 1926–27 from French, Spanish, and Portuguese archives. The locations of libraries that have parts or all of the Karpinski series are given in Karpinski, *Bibliography of the Printed Maps of Michigan, 1804–1880* (Lansing: Michigan Historical Commission, 1931), p. 24.

The Hulbert and Karpinski reproductions are not accompanied by annotation or comment. A useful documented selection of seventy-five maps is found in Emerson D. Fite and Archibald Freeman, *A Book of Old Maps Delineating American History from the Earliest Days Down to the Close of the Revolutionary War* (Cambridge, Mass.: Harvard University Press, 1926; reprinted by Dover Publications, New York, 1969, with reproductions made from new and improved plates but still often illegible).

The largest collection of photocopies of American maps, made from originals in European and American libraries (neither published nor fully catalogued), is in the Geography and Map Division of the Library of Congress.

Atlases

The atlases of the geographers and commercial mapmakers, who made use of earlier surveys and reports to which they gained access from private sources or from official archives, are a useful indication of cartographical progress during the eighteenth century. The first edition of *The English Pilot, Fourth Book* appeared in London in 1689; its charts of the North American coast were issued in numerous editions until 1794, with occasional additions of new maps and revisions of the plates. Herman Moll was the most diligent mapmaker in London in the first third of the century in gathering and incorporating the latest information, English and continental, for his various atlases, which included *The World De-*

scribed (London, 1709–20), and *The Atlas Minor* (London, 1729). Henry Popple's "A Map of the British Empire in America with the French and Spanish Settlements adjacent thereto" (London, 1733) is the first large-scale printed map of eastern North America, often bound in twenty-one sheets in the form of an atlas. Popple's map includes insets of twenty-two cities, harbors, ports, and islands. Another great wall map is John Mitchell's "A Map of the British and French Dominions in North America" (London, 1755). Mitchell made excellent use of much unpublished information from private and official sources; his is the best general map of the North American colonies for the middle of the century. The year 1755 and after saw the publication of several large-scale maps of the different colonies and provinces by Thomas Jefferys; they were later incorporated in his *A General Topography of North America* (London, 1768). Many of these same maps, with additional items, are found in Jefferys's posthumous *American Atlas* (London: R. Sayer and J. Bennett, 1775). Especially rich in revolutionary war maps are William Faden's *North American Atlas* (London: W. Faden, 1777) and his *Atlas of Battles of the American Revolution* (London, ca. 1793).

In 1774 and after Des Barres produced a series of over two hundred and fifty magnificent charts and sketches of the North American Atlantic coast that are found in *The Atlantic Neptune*, published by the British Admiralty. No set of volumes is complete and no two sets alike, since each was selected for a particular purpose, mission, or assignment. Barre Publishers, of Barre, Massachusetts, has recently published a facsimile set of these charts.

RESOURCES OF THE BRITISH MUSEUM

In the British Museum is an unsurpassed collection of North American plans, charts, and maps. Many of these are in King George III's Topographical Collection; the king, himself an avid map collector, was in a position to select what he wanted. Other maps of significance are among the additional manuscripts that have come to the museum through the years from various sources. In 1968 it acquired the map collection of the Royal United Services Institution. This is the largest single purchase of historical maps in the museum's history: 3,500 maps, of which 988 are manuscript. The map collection had been built up since the founding of R.U.S.I. in 1831 by deposits from officers serving in the field and from government departments, but chiefly from bequests of complete collections made by officers or their heirs. One, given by the third baron Amherst about 1861, consists of some 250 maps, of which 126 are manuscript. Most of these were gathered by his great-uncle, Field Marshal Jeffrey Amherst, later first baron, while he was commander-in-chief of British forces in North America, 1758–63. Among them are plans and maps of the mighty fortress of Louisbourg, on Cape Breton Island, which Amherst was sent to capture by Pitt in 1758; maps of Ticonderoga and Crown Point, where he was also victorious; beautifully drawn colored maps by young artillery engineers of the routes to Montreal; and French maps, possibly taken at Montreal after its fall. Also in the Royal United Services Institution Collection is the bequest of Colonel Sir Augustus Simon Frazer, which he made about 1835; only thirty-two manuscript maps are in the Frazer collection but among them are maps of significant historical interest. Sir Augustus was born in 1776 and did not serve in America; where he acquired his American maps is not known. He was in charge of the Horse Guard Artillery at Woolwich; he became director of the Royal Laboratory there and may have gathered his maps then.

Another smaller group of R.U.S.I. maps has the designation R.A.D.; it contains, together with nineteenth-century maps, pre-revolutionary maps of North American and West Indian interest. The donor of these was R. A. Davenport (1777?–1852), a voluminous writer of historical, geographical, and biographical works.

The largest group of manuscript maps in the collection, 482, are in the bequest of Lieutenant-General Sir Harry David Jones, G.C.B., R.E. This "HJ" Collection, ranging from 1673 to 1806, consists of maps relating to the War of the Spanish Succession, Prussian campaigns in Bohemia in the mid-eighteenth century, and the Russian and Austrian campaigns against the Turks that include an important collection of maps of actions at Belgrade in 1717 and 1735. Jones was attached as a commissioner to the Prussian army of occupation in France, 1816–18; it is probable that in some way he acquired the maps from a Prussian archive directly or from a French source to which it had been removed during the Napoleonic wars. Jones was governor of the Royal and Military Staff Colleges (Sandhurst) from 1856 until his death in 1866. See Whitworth Porter, *History of the Corps of Royal Engineers,* 3 vols. (London: Longmans, Green, 1889–1915), 2:445–48.

The printed and engraved maps in the various R.U.S.I. bequests contain numerous rare items but do not equal in importance the manuscripts; together they comprised at the time of their purchase the largest private collection of historical maps then in Great Britain. Before their acquisition the writer examined the R.U.S.I. Collection in February 1968 for the British Museum as consultant for the American maps. The maps were housed on the fourth story of the R.U.S.I. building, adjoining the banqueting hall on Whitehall, in several chests, a large wardrobe, and over thirty wooden drawers. They had been catalogued in 1867 and again in 1910. The maps now bear British Museum Add. mss. numbers.

Supplementing and often duplicating military maps of the third quarter of the eighteenth century in the British Museum and the Public

Record Office is a large collection of manuscript and printed maps in the Duke of Cumberland Archives in the private Royal Library in Windsor Castle. The Duke of Cumberland (1721–1765) was the captain general of the British army from 1745 to 1757. The major part of the map collection relates to military activities on the continent of Europe and in India; maps of the early years of the French and Indian War form the largest part of the American military maps, although the collection extends to American revolutionary years, some time after the Duke of Cumberland's death. The hand-written map catalogue lists the entire collection alphabetically, not geographically.

REGIONAL BIBLIOGRAPHIES AND MAJOR AMERICAN MAP COLLECTIONS
Southeast

Louis C. Karpinski, editor, *Early Maps of Carolina and Adjoining Regions . . .* from the Collection of Henry P. Kendall (2d ed.; Charleston, S. C.: privately printed, 1937), is a catalogue of the Kendall Collection, now housed in the South Caroliniana Room of the University of South Carolina. It includes world and North American maps and extends through the nineteenth century. William P. Cumming, *The Southeast in Early Maps* (2d ed.; Chapel Hill: University of North Carolina Press, 1962) is an annotated checklist of printed and manuscript regional and local maps of southeastern North America south of Virginia during the colonial period.

Florida. Woodbury Lowery, *The Lowery Collection: A Descriptive List of Maps of the Spanish Possessions within the Present Limits of the United States, 1502–1820* (Washington, D.C.: Government Printing Office, 1912), has extensive and valuable bibliographical notes by Phillips on Lowery's list, which after 1700 emphasizes Spanish maps of the Florida peninsula.

Georgia and Alabama. "The Rucker Agee Collection of the Birmingham Public Library" (Birmingham: Birmingham Public Library, 1964), is a mimeographed catalogue listing Agee's cartographic collection. It now has over two thousand sheet maps and a good collection of atlases and related books.

"Georgia and Southeastern maps from the Ivan Allen Collection in the Emory University Library June 19, 1961" is a mimeographed catalogue listing original maps and photocopies of maps in the Emory University Library, Decatur, Georgia.

A small collection of rare eighteenth-century maps of the southeast, including some valuable manuscript maps, is in the Ilah Dunlap Little Memorial Library, University of Georgia, Athens; many are listed with annotations in the *Catalogue of the Wymberley Jones DeRenne Georgia*

Library, L. L. Mackall, cataloguer, and Azalea Clizbee, compiler, 3 vols. (Wormsloe, Georgia: privately printed, 1931).

South Carolina. Mentioned above is the catalogue by L. C. Karpinski of the Kendall Collection, which is now housed in the South Caroliniana Library of the University of South Carolina in Columbia.

A list of plans and charts of Charleston in American and British archives is in Helen G. McCormack, "A Catalogue of Maps of Charleston," *Year Book of 1944: City of Charleston, S. C.* (Charleston, S. C., 1947 [1948]).

The South Carolina Department of Archives in Columbia has several score of volumes of plats and surveys of South Carolina that go back to the early years of the province and are largely unexplored.

North Carolina. A list of North Carolina maps extending through the nineteenth century is in F. B. Laney and K. H. Wood, *Bibliography of North Carolina Geology, Minerology and Geography With a List of Maps*, North Carolina Geological and Economic Survey, Bulletin 18 (Raleigh, N. C., 1908), Part 2, "List of Maps of North Carolina," pp. 269–362.

A selected list of maps with large-scale reproductions, accompanied by cartographic and historical notes, is in William P. Cumming, *North Carolina in Maps* (Raleigh, N. C.: State Department of Archives and History, 1966).

Good collections of maps of the southeast are in the North Carolina Room, University of North Carolina Library, in Chapel Hill, and in the North Carolina State Department of Archives and History in Raleigh.

Middle Atlantic

No general study of maps of this area exists, although facsimiles and comments are scattered through the volumes of Winsor's *America*.

Virginia. The best general listing is in Earl G. Swem, *Maps Relating to Virginia in the Virginia State Library and other Departments of the Commonwealth, with the 17th and 18th Century Atlas-Maps in the Library of Congress* (Richmond: Davis Bottom [superintendent of public printing], 1914), pp. 33–263. A cartobibliography of printed maps is in preparation by Professor Coolie Verner, who has already examined several specific phases of the subject: *A Carto-bibliographical Study of the English Pilot, The Fourth Book, with Special Reference to the Charts of Virginia* (Charlottesville: University of Virginia Press, 1960): "Smith's Virginia and Its Derivatives," *The Map Collectors' Series, No. 45* (London: Map Collectors' Circle, 1968); "The Fry and Jefferson Map," *Imago Mundi* 21 (1967):70–94.

Careful studies of selected Virginia maps are found in Philip Lee Phillips, "Virginia Cartography," *Smithsonian Miscellaneous Collections*, no. 37 (Washington: Smithsonian Institution, 1896); and Fairfax Harrison, *Landmarks of Old Prince William*, 2 vols. (Richmond: Dietz Press, 1924).

The University of Virginia Library has a number of rare Virginia maps, including the first edition of the Fry-Jefferson Virginia 1751 [1753]. The Mariners Museum, Newport News, with its emphasis on maritime history, has acquired since 1930 a good collection of atlases and charts. It lists over five thousand maps in its *Catalog of Maps, Ships' Papers and Logbooks* (Boston: G. K. Hall, 1964).

Washington, D. C. The Library of Congress, with over three million maps and thirty thousand atlases, has the largest collection of American maps (including eighteenth-century British maps) in any library. Philip Lee Phillips's *List of Maps of America* (1901) and *List of Atlases* (1909–   ), already mentioned, are indispensable tools for the student. The single-sheet map collection of the library is particularly valuable because it includes photocopies of distinctive maps from other repositories as well as its own printed and manuscript originals. It has a number of special collections containing eighteenth-century manuscript maps of America; the following deserve special mention.

The William Faden Collection of French and Indian War and Revolutionary War Maps (101 maps). Many of these were the original works that Faden used for the engraved maps in his atlases. They are listed in [Edward E. Hale, compiler] *Catalogue of a Curious and Valuable Collection of Original Maps and Plans . . . the Collection of William Faden, the "King's Geographer"* (Boston: J. Wilson & Son, 1852).

Admiral Lord Richard Howe Collection of Manuscript Maps (72 maps) covers portions of the Atlantic coast, the West Indies, and the Philippine Islands. Supplementary to these are the Comte de Rochambeau Collection of Revolutionary War Maps (66 maps) and the Ozanne Collection of Manuscript Views and Maps of the Revolutionary War (23 maps and views).

The Library of Congress also has a bibliography of cartography that consists of over forty thousand cards, with annotations, quotations, and references to mapmakers, mapmaking, and the literature about maps, now published as *The Bibliography of Cartography* (Boston: G. K. Hall, 1973).

Recent publications give helpful information on the location of collections and of individual maps of the Revolution. Walter W. Ristow, "Maps of the American Revolution: A Preliminary Survey," *The Quarterly Journal of the Library of Congress* 28 (July 1971):196–215, has an excellent general discussion of the subject. Peter J. Guthorn, *American Maps and Map Makers of the Revolution* (Monmouth Beach, N. J.: P. Freneau Press, 1966), and Peter J. Guthorn, *British Maps of the American Revolution* (Monmouth Beach, N. J.: P. Freneau Press, 1972). David S. Clark, *Index to Maps of the American Revolution in Books and Periodicals* (Washington, D. C.: privately published, 1969), is a mimeographed list of maps found in later publications, chiefly after 1800.

Maryland. The most useful work on Maryland maps of this period is Edward B. Mathews, *The Maps and Map-Makers of Maryland*, Maryland Geological Survey, special publication, vol. 2, part 3b (Baltimore: Johns Hopkins Press, 1898).

Pennsylvania. A good general survey of Pennsylvania maps is Hazel Shields Garrison, "Cartography of Pennsylvania before 1800," *Pennsylvania Magazine of History and Biography* 59 (1935):225–83. Several scholarly works on Pennsylvania mapmakers include helpful cartographic information beyond the bounds of the province and the work of the individual mapmaker. See, for example, Lawrence H. Gipson, *Lewis Evans. With Evans' "A Brief Account of Pennsylvania" (1753). Also facsimiles of his Essays, Numbers I & II* (Philadelphia: Historical Society of Philadelphia, 1939), thirty-one facsimile maps. Thomas Pownall, *A Topographical Description of the Dominions of the United States of America,* ed. Lois Mulkearn, with extensive notes and bibliography (Pittsburgh: University of Pittsburgh Press, 1949). Howard N. Eavenson, *Map Maker and Indian Traders* (Pittsburgh: University of Pittsburgh Press, 1949), an account of John Patten, maker of "The Trader's Map" in the Library of Congress. Walter Klinefelter, "Surveyor General Thomas Holme's Map of the Improved Part of the Province of Pennsylvania," *Winterthur Portfolio 6* (Charlottesville: University Press of Virginia, 1970).

A helpful local bibliography is Philip Lee Phillips, *A Descriptive List of Maps and Views of Philadelphia in the Library of Congress 1683–1865,* Special Publication No. 2 (Philadelphia: Geographical Society of Philadelphia, 1926).

New York. Winsor's *America* is still the best source of cartographical information for the province, with facsimiles and comments especially useful for the French and Indian War and the revolutionary war periods. One of the most complete and scholarly works of its kind for any part of North America is I. N. P. Stokes, *The Iconography of Manhattan Island, 1498–1909,* 6 vols. (New York: Robert H. Dodd, 1915–28). A good index of maps is in Daniel C. Haskell, ed. *Manhattan Maps: A Co-operative List* (New York: New York Public Library, 1931).

In New York City, the New York Public Library, enriched by the Stokes and other collections, the American Geographical Society, and the New-York Historical Society, with manuscript maps of the revolutionary war period, have rare printed and manuscript maps.

New England
One of the chief desiderata in the cartobibliography of colonial North America is a study of the New England region. Here again Winsor's *America* provides a helpful start, with numerous reproductions and bibliographical notes scattered through several volumes.

Connecticut. Edmund B. Thompson, *Maps of Connecticut Before the Year 1800* (Windham, Conn.: Hawthorn House, 1940), is one of the best examples, in format and documentation, of what can be produced in listing the maps of a colonial province. Still useful is Clarence Winthrop Bowen, *The Boundary Disputes of Connecticut* (Boston: Osgood & Co., 1882).

The map collection in the Yale University Library is notable both in the number and quality of its holdings.

Rhode Island. Howard Millar Chapin, *Cartography of Rhode Island. Contributions to Rhode Island Bibliography,* no. 3 (Providence: Preston & Rounds, 1915). This list attempts completeness neither in the maps listed nor in their documentation.

One of the great map collections in the United States is in the John Carter Brown Library, Providence, both in manuscript and in rare printed maps of the Americas before 1800. Although it has no printed map catalogue, its important recent acquisitions are described in its annual reports that began in 1911. The *Annual Reports 1901–66* (8 vols.) of the John Carter Brown Library have now been reprinted by Atheneum Publishers.

Massachusetts. Winsor's *America* has numerous reproductions and critical notes on colonial Massachusetts maps; for the Boston area a helpful work with numerous reproductions and plans of Boston and vicinity is Justin Winsor, *Memorial History of Boston,* 3 vols. (Boston: J. R. Osgood, 1880–83). Two lists of maps of Boston are by the Boston, Massachusetts, Engineering Department, *A List of Maps of Boston Published between 1600 and 1903* (reprint, Appendix 1, Annual Report of City Engineer; Boston: Municipal Printing Office, 1903); and by Philip L. Phillips, *A Descriptive List of Maps and Views of Boston in the Library of Congress, 1630–1865* (Washington, 1922), a 275-page typewritten list of which there are copies in the Library of Congress and in the Justin Winsor Room of the Harvard College Library.

The Harvard College Library has a large collection of maps. The Winsor Memorial Map Room has many atlases and maps, including the rare William Douglass New England (1753); the Houghton Library has J. G. W. De Brahm's manuscript *History of the Three Provinces South Carolina Georgia and East Florida* (1772) with fourteen maps, and eleven manuscript maps of the Revolutionary War period from the Jared Sparks Collection (other Sparks maps are in the Cornell University Library). The only printed catalogue of the map collection is *A Catalogue of the Maps and Charts in the Library of Harvard University in Cambridge, Massachusetts* (Cambridge, Mass.: E. W. Metcalf and Co., 1831).

The Boston Public Library, the Athenaeum, and the Massachusetts Historical Society have valuable though not large collections of maps.

Maine. A brief list of maps is in Edgar Crosby Smith, *Maps of the State of Maine; a bibliography of the maps of the State of Maine* (Bangor: C. H. Glass & Co., 1903).

In Appendix B is a list of Bernard maps, which includes some manuscript maps of Maine.

New Hampshire and Vermont. David A. Cobb, "Vermont Maps Prior to 1900: An Annotated Cartobibliography," *Vermont History: The Proceedings of the Vermont Historical Society* 39 (1971):169–317.

Canada

A good account of maps and mapmakers of Canada is in Don W. Thomson, *Men and Meridians.* Vol. 1: Prior to 1867 (Ottawa: Queen's Printer, 1966). Reproductions of significant maps, chiefly French, with accompanying annotations, are in Marcel Trudel, *An Atlas of New France* (Ottawa: Laval University Press, 1968).

The Public Archives of Canada in Ottawa has an excellent collection of manuscripts and printed maps and atlases, with photocopies of maps and charts in other repositories.

Trans-Allegheny and the Mississippi Valley

A fine study, with full documentation, of selected maps illustrating the westward British push across the mountains is Lloyd Arnold Brown, *Early Maps of the Ohio Valley* (Pittsburgh: University of Pittsburgh Press, 1959). A list with occasional annotations is C. C. Baldwin, "Early Maps of Ohio and the West," *Western Reserve and Northern Ohio Historical Society,* tract no. 25 (Cleveland, 1875), pp. 7–25. Louis C. Karpinski, *Bibliography of the Printed Maps of Michigan 1804–1880 . . . with . . . An Historical Atlas of the Great Lakes and Michigan* (Lansing: Michigan Historical Commission, 1931), combines a bibliography with useful notes and an atlas. The list of maps before 1804 is selective but includes a wider area than the title indicates.

A cartobibliography of maps showing the central Mississippi Valley

region, which, however, is far from exhaustive, is Willard Rouse Jillson, *A Checklist of Early Maps of Kentucky (1673–1825)* (Frankfort, Ky.: Roberts Printing Co., 1949).

The William L. Clements Library, Ann Arbor, Michigan, has a large collection of atlases and separate printed maps, listed in Douglas W. Marshall, ed., *Research Catalog of Maps of America to 1860 in the William L. Clements Library,* 4 vols. (Boston: G. K. Hall, 1972). Its manuscript maps of the French and Indian War and the revolutionary war periods are unrivaled in this country. Christian Brun, *Guide to the Manuscript Maps in the William L. Clements Library* (Ann Arbor: University of Michigan Press, 1959), lists over eight hundred manuscript maps and plans from the Sir Henry Clinton, General Thomas Gage, Lord George Germain, Earl of Shelburne, and other collections. A Murray map of the Saint Lawrence River (1761) in eighteen sheets and a Stuart-Purcell map (ca. 1775) are notable.

The Newberry Library in Chicago is another of the great repositories of early maps, charts, and atlases, both printed and manuscript, in this country. Although the Newberry is especially rich in sixteenth-century printed maps, eighteenth-century manuscript maps comprise a major part of the items listed in Clara A. Smith, *List of Manuscript Maps in the Edward E. Ayer Collection* (Chicago: Newberry Library, 1927). The Stuart-Purcell manuscript map of the Southern District (1775) is one of the great maps of the Southeast. Thomas Mante, *The History of the Late War in North-America* (London, 1772), is extra-illustrated with manuscript maps and notes on the French and Indian War. Among the rare printed maps are a first state of Farrer's "Virginia" (1651) and an even rarer fourth state of Moll's "Dominions of the King of Great Britain . . ." (1715). For a partial list of the library's holdings see the *Dictionary Catalog of the Edward E. Ayer Collection of Americana,* 16 vols. and 3-vol. supplement (Boston: G. K. Hall, 1961).

The Far West

Original British contributions to the cartography of the West are minor until the last quarter of the century. Two works are preeminent for the area: Carl I. Wheat, *Mapping the Trans-Mississippi West,* Vol. 1: 1540–1804 (San Francisco: Institute of Historical Cartography, 1957); and Henry R. Wagner, *The Cartography of the Northwest Coast of America to the Year 1800,* 2 vols. (Berkeley: University of California Press, 1937). Among more specialized studies may be mentioned R. V. Tooley, "California as an Island," *Map Collectors' Series, No. 8* (London: Map Collectors' Circle, 1964).

The Henry E. Huntington Library and Art Gallery in San Marino, California, has the best collection of printed and manuscript British maps on the West Coast. These include manuscript maps and plans of the period of the French and Indian War from the collections of Lord Loudoun and General Abercromby, and of General Braddock's maps captured by the French at Fort Duquesne. For the Revolution, the library has the Simcoe Papers, with maps made by officers in the Queen's Rangers under Lieutenant-Colonel John G. Simcoe. These maps, like those in the Public Record Office, often accompany correspondence that throws light on their provenance.

Many university libraries, historical societies, and state archives not mentioned in this essay have collections of maps and atlases of the colonial period. A helpful and comprehensive listing of the cartographic resources of libraries in North America, though often without detailed indication of special resources, is given in *Map Collections in the United States and Canada: A Directory: Second Edition*. Compiled by the Directory Revision Committee: David K. Carrington, Chairman (New York: Special Libraries Association, 1970).

The need and the opportunity for further examination of important individual maps as well as for historical and cartobibliographical studies in regional maps of the colonial period are clearly evident.

Index